Fluid Mechanics Experiments

Synthesis Lectures on Mechanical Engineering

Synthesis Lectures on Mechanical Engineering series publishes 60–150 page publications pertaining to this diverse discipline of mechanical engineering. The series presents Lectures written for an audience of researchers, industry engineers, undergraduate and graduate students.

Additional Synthesis series will be developed covering key areas within mechanical engineering.

Fluid Mechanics Experiments
Robabeh Jazaei
2020

Introduction to Deep Learning for Engineers: Using Python and Google Cloud Platform
Tariq M. Arif
July 2020

Towards Analytical Chaotic Evolutions in Brusselators
Albert C.J. Luo and Siyu Guo
2020

Modeling and Simulation of Nanofluid Flow Problems
Snehashi Chakraverty and Uddhaba Biswal
2020

Modeling and Simulation of Mechatronic Systems using Simscape
Shuvra Das
2020

Automatic Flight Control Systems
Mohammad Sadraey
2020

Bifurcation Dynamics of a Damped Parametric Pendulum
Yu Guo and Albert C.J. Luo
2019

Mathematical Magnetohydrodynamics
Nikolas Xiros
2018

Design Engineering Journey
Ramana M. Pidaparti
2018

Introduction to Kinematics and Dynamics of Machinery
Cho W. S. To
2017

Microcontroller Education: Do it Yourself, Reinvent the Wheel, Code to Learn
Dimosthenis E. Bolanakis
2017

Solving Practical Engineering Mechanics Problems: Statics
Sayavur I. Bakhtiyarov
2017

Unmanned Aircraft Design: A Review of Fundamentals
Mohammad Sadraey
2017

Introduction to Refrigeration and Air Conditioning Systems: Theory and Applications
Allan Kirkpatrick
2017

Resistance Spot Welding: Fundamentals and Applications for the Automotive Industry
Menachem Kimchi and David H. Phillips
2017

MEMS Barometers Toward Vertical Position Detection: Background Theory, System Prototyping, and Measurement Analysis
Dimosthenis E. Bolanakis
2017

Engineering Finite Element Analysis
Ramana M. Pidarti
2017

Fluid Mechanics Experiments
Robabeh Jazaei

ISBN: 978-3-031-79672-2 print
ISBN: 978-3-031-79673-9 ebook
ISBN: 978-3-031-79674-6 hardcover

DOI 10.1007/978-3-031-79673-9

A Publication in the Springer series
SYNTHESIS LECTURES ON MECHANICAL ENGINEERING
Lecture #29

Series ISSN   2573-3168   Print   2573-3176   Electronic

Fluid Mechanics Experiments

Robabeh Jazaei
Slippery Rock University of Pennsylvania

SYNTHESIS LECTURES ON MECHANICAL ENGINEERING #29

ABSTRACT

Fluid mechanics is one of the most challenging undergraduate courses for engineering students. The fluid mechanics lab facilitates students' learning in a hands-on environment. The primary objective of this book is to provide a graphical lab manual for the fluid mechanics laboratory. The manual is divided into six chapters to cover the main topics of undergraduate-level fluid mechanics. Chapter 1 begins with an overview of laboratory objectives and the introduction of technical laboratory report content. In Chapter 1, error analysis is discussed by providing examples. In Chapter 2, fluid properties including viscosity, density, temperature, specific weight, and specific gravity are discussed. Chapter 3 revolves around the fluid statics include pressure measurement using piezometers and manometers. Additionally, hydrostatic pressure on the submerged plane and curved surfaces as well as buoyancy and Archimedes' Principle are examined in Chapter 3. In Chapter 4, several core concepts of fluid dynamics are discussed. This chapter begins with defining a control system based on which momentum analysis of the flow system is explained. The rest of the chapter is allotted to the force acting on a control system, the linear momentum equation, and the energy equation. Chapter 4 also covers the hydraulic grade line and energy grade line experiment. The effect of orifice and changing cross-sectional area by using Bernoulli's' equation is presented in Chapter 4. The application of the siphon is extended from Chapter 4 by applying Bernoulli's' equation. The last two chapters cover various topics in both internal and external flows which are of great importance in engineering design. Chapter 5 deals with internal flow including Reynolds number, flow classification, flow rate measurement, and velocity profile. The last experiment in Chapter 5 is devoted to a deep understanding of internal flow concepts in a piping system. In this experiment, students learn how to measure minor and major head losses as well as the impact of piping materials on the hydrodynamics behavior of the flow. Finally, open channels, weirs, specific energy, and flow classification, hydraulic jump, and sluice gate experiments are covered in Chapter 6.

KEYWORDS

laboratory manual, fluid mechanics, fluid statics, fluid dynamics, internal flow, external flow

To the Students

With the hope that this work will stimulate an interest in the Fluid Mechanics Laboratory, report preparation, and provide graphical guidance for some of its experiments.

Contents

Acknowledgments

The author would like to acknowledge with appreciation the numerous valuable comments, suggestions, constructive criticisms, and praise from faculty at the University of Wisconsin-Platteville and the University of Nevada, Las Vegas. Many thanks are extended to all my colleagues and my students at the University of Wisconsin-Platteville who provided plenty of feedback over the last two years of teaching fluid mechanics and its laboratory. I also would like to thank Ms. Rebecca Messer for her assistance in editing the initial draft of the book.

I deeply and sincerely appreciate the continual encouragement and support of my family. I would greatly appreciate hearing from you if you have any comments, suggestions, or problems related to the manual.

Robabeh Jazaei, Ph.D.
July 2020

CHAPTER 1

How to Write a Technical Report

1.1 INTRODUCTION

Lab reports are an important part of laboratory courses in all engineering disciplines because accurate wording is as important as accurate calculations. The objective of writing a lab report is to learn how to effectively communicate with others (i.e. clients, peer engineers, etc.) and present the result of an experiment in detail. Often a good lab report that is concise and to the point, supported with accurate data and analysis, will secure more work in students' professional engineering career.

1.1.1 WHAT IS THE PURPOSE OF A LABORATORY REPORT?

There are two versions of a laboratory report including laboratory memorandum or simply "lab memo" and laboratory technical report. In fact, laboratory memo is a short version of a technical report. Sometimes, students find the write up of a lab memo more difficult than a technical report because a memo contains all the same components of a lab report in a simplified manner. In this section, the purpose of a technical report and graphical step-by-step of writing a lab report are presented.

Each experiment is designed to answer a question or test a hypothesis. However, raw data cannot be supported or interpreted by itself. The lab report is a detailed technical documentation to describe the objectives, methodology, and results of the experiment. The three main objectives of a lab report are to address why the experiment was done (objectives), how the experiment was performed (procedure), and what was determined through the experiment (results). The objective, procedure, and results must cover the following content.

- The objectives of the experiment should be stated clearly.

- The methodology should be detailed for another individual to be able to replicate the experiment and obtain the same results.

- The results should be discussed clearly. Graphs and tables help readers to visually understand the results.

- If necessary, provide some recommendations or explain why the results are not the same as theoretical values. Discuss the uncertainty of the results.

1.1.2   LAB REPORT FLOWCHART AND STEP TO WRITE A LAB REPORT

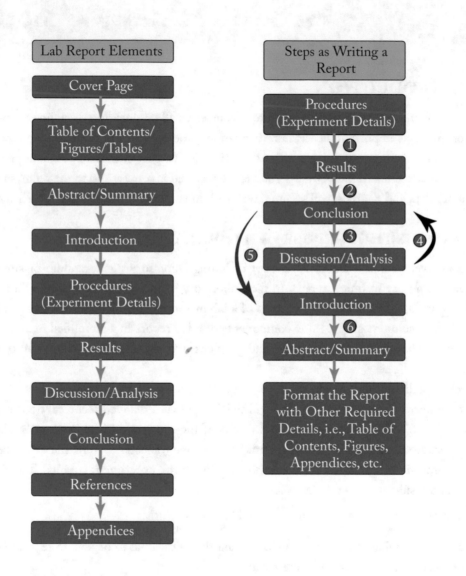

1.2   COVER PAGE

University Logo

Title of the Experiment

Lab

Course and Lab Section#

Prepared For:

Your Instructor/ Supervisor/Client

Department of Civil and Environmental Engineering

Name of the University

Prepared By: Student's name

Team Members' name

Lab Performed: Date the experiment was done

Lab Submitted: Date the report is due

1.3   TABLE OF CONTENTS/FIGURES/TABLES

In the table of contents in the lab report, the list of major and minor contents is provided. Likewise, the table of tables and table of figures are created.

1.4   ABSTRACT/SUMMARY

The abstract is the first section of your lab report that the audience will read, and it should stand by itself. Sometimes, the client or supervisor will not have time to read the whole report, instead they read the abstract. If the reader seeks more information, they will go over details in the report. The abstract should be written after you have completed writing the report, so you are familiar with all major findings.

- The abstract is a summary of the experiments including objectives, methodology, results, and/or recommendation.

- Use passive voice, i.e., "the mass was measured" instead of "I measured the mass" or "We measured the mass."

- For a smooth transition, it is recommended to begin with one or two statements as an introduction.

- Help the reader quickly understand the purpose of the report by clearly stating the objectives of the experiment.

- Quantify the results and provide uncertainty if applicable, i.e., "The hydrostatic force increased by 20%" instead of "The hydrostatic force increased significantly."

- Do not refer to a table or graph in the abstract. Keep in mind the abstract should be able to stand by itself.

- Be concise and to the point (maximum of 250 words).

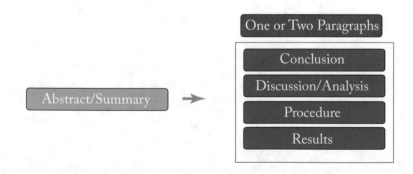

1.4.1   HOW TO WRITE A GOOD ABSTRACT OR SUMMARY

Use the structures as shown in the following example (Impact of Jet or Momentum Lab), including introduction, objective(s), brief procedure, and results. It is important that the abstract stands by itself. The same technique can be used to write abstract for any research paper in graduate school or in the engineering professions. Sometimes a good abstract works as a hook to encourage the reader to explore the written paper. See the color-coded example.

Example of a Good Abstract—Momentum Lab

"Momentum is used for calculation propulsion in airplane, ships, and so on. Momentum refers to the quantity of motion an object has and is a function of the mass and velocity of that object. The objective of this experiment was to gain a better understanding of the force of a jet-stream of water on an object. This was achieved by using a jet apparatus on a flat, hemispherical, and 120° vane and using calibrated weights to balance out the apparatus. The value of weight, along with the flow and other given constants were used to determine the force acting on the plate. The results indicated that the 120° vane required the least amount of force due to the angle associated with the vane. Because of this angle, the stream has vertical and horizontal force components but only the vertical component contributed to the force, resulting in a force with less magnitude than the others. From each experiment, it was found that the percent error was 18.27% for the 180° vane, 11.17% for the 120° vane, and 14.68% for the 90° vane."

Color Code:

Introduction

Objective

Procedure

Results

Abstract Word Count: 181

1.5   INTRODUCTION

The purpose of an introduction is to motivate the reader with enough theoretical or historical background to further understand why this experiment was performed. Do not include the summary of results and conclusions in the introduction section. Governing equations can be included in either the introduction or in the sample calculations listed in the appendices. See the color-coded example.

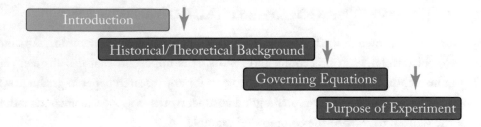

Example of a Good Introduction– Momentum Lab

"Hydraulic machinery is used in countless applications today. From heavy equipment, such as bulldozers, to the automobile's automatic transmission and braking system, to a simple barber chair, hydraulic machines are widespread. One way to gain an understanding of hydraulic machinery is to study the force on an object from the impact of a jet. When high water pressure flows through a nozzle and hits a vane, the energy from the water pressure is converted to kinetic energy. The impact from the jet apparatus determines the force on the plates on which it is aimed. The objective of this lab was to study these forces."

Color Code:
Application and background
Propose of experiment

1.6   PROCEDURE

The procedure (methodology) should provide enough detail that the reader(s) can repeat the experiment to verify the results. The procedure should include:

- a description of the apparatus. Draw, sketch, or add pictures so the reader can visualize the apparatus and the procedure;

- a record of the measurement made. Provide the manufacturer model, and uncertainty of the apparatus, if applicable;

- a list of any precautions you took to ensure accuracy of the experiment; and

- a mention of unexpected difficulties and how you overcame them.

Do not use bulleted statements.

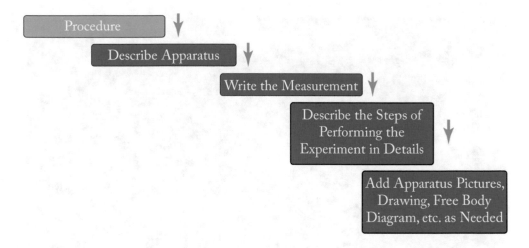

Example of a Good Procedure—Momentum Lab

For this experiment, a jet apparatus shown in Figure 1.1 was used to determine the force acting on a series of plates. The jet was positioned on the hydraulic bench and leveled by adjusting the three leveling feet until the spirit level, located on top of the device, was centered. The top of the apparatus was removed, the 90° vane was placed inside the apparatus, and the device was securely closed by screwing on three wing-nuts. The pointer was aligned with the weight carrier on top of the apparatus prior to initiating the water jet. The pump was turned on and adjusted to a desired, steady flow. Appropriate masses were then added to the weight carrier. The flow of the water was adjusted to rebalance the pointer. The masses added to the weight carrier were recorded in Table 1.1 and ranged from 50–300 grams. The water flowed through the nozzle and hit the vane before exiting the cylinder through a channel that fed a drain to the piezometer. The volume of water in the piezometer and the corresponding time were recorded in Table 1.1 to determine the flow rate ($Q=V/t$, litter/second). The flow rate and mass were then subsequently recorded. The weight and accompanying flow-rate were adjusted several times and the recording process repeated. After completing four trials for the flat plate, the hydraulic bench valve was closed, and the pump was shut off to allow the apparatus to drain. The procedure was repeated over a series of four trials for each plate type. The flat plate was then replaced with the hemispherical and conical plates. The results of the respective trials were recorded in Tables 1.2 and 1.3.

… continued…

Figure 1.1: Impact of a jet apparatus detail.

Color Code:
Describe the apparatus
State the step-by-step procedure
Write measurements
Precautious for accuracy

1.7   RESULTS

All the measurements which made during the experiment along with some indication of the uncertainty of each measurement should be recorded, if applicable. Use illustration to present the results in the form of figures and tables when possible.

- Start with the results narrative.

- All figures are followed by a caption written below each figure and ended with a period.

- All figures and tables should be discussed within the text.

- If an equation is used, describe each variable along with the proper unit.

- The experimental data should be compared to theoretical predictions and calculations.

- Error analysis should be included.

Example of a Figure and Table with Correct Citations –Hydrostatics Lab

a. Figure:

The caption for figures should be below the picture or graph and only the first word is capitalized. The figure caption is followed by a period, as shown in Figure 1.2. The graphs or figures also should stand by itself. Thus, use a proper caption and clear labels for each axis.

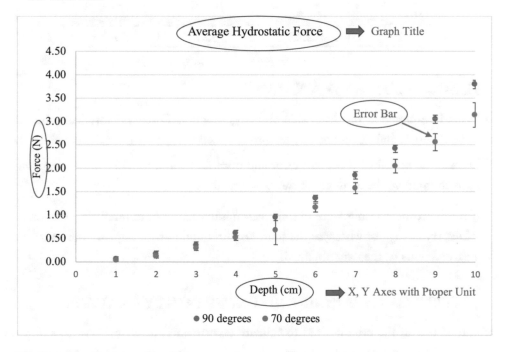

Figure 1.2: Relationship of average hydrostatic force as a function of depth with error bars.

b.   Table:

All tables include a caption which is located directly above the table. The caption is capitalized but it is not followed by a period. Use proper SI units unless stated otherwise. The values should be in center of each cell (Table 1.1).

Table 1.1: Theoretical and Experimental Hydrostatic Forces for 90°

Depth of Water (cm)	Theoretical F_R (N)	Experimental F_R (N)	Percent Error
1	0.25	0.16	37.0%
2	0.50	0.29	41.1%
3	0.74	0.49	33.8%
4	0.99	0.74	25.8%
5	1.24	1.08	12.6%
6	1.49	1.49	0.1%
7	1.74	2.00	15.3%
8	1.99	2.57	29.4%
9	2.23	3.22	44.2%
10	2.48	3.97	59.7%

1.8   DISCUSSION AND ANALYSIS

The results are interpreted in the discussion or analysis section. Discuss theoretical values in comparison to the experimental values and any reasons the results may vary from the anticipated results.

- Discuss the meaning of percent error and uncertainty and provide any recommendations to improve the accuracy of the experiment.

- If the experiment results were in some way unsatisfactory or different from theoretical values, try to suggest what was the source of error and how to improve the accuracy of measurements or the data obtained.

1.8.1   STATISTICAL ANALYSIS AND UNCERTAINTY ANALYSIS

Repeatability is one of the criteria used to validate experimental investigations. In short, repeatability is the ability of a test method to reproduce similar results in successive measurements. In other words, an experiment is considered repeatable if the variation of the results does not exceed a predetermined value when an observer repeats the test several times using the same measuring devices and under the same conditions over a short period of time.

Repeatability eliminates the bias error in the cost of collecting more data and processing them. Repeating an experiment results in collecting a set of discrete data that need to be classified and interpreted systematically to produce scientifically reliable results.

Statistics is the branch of mathematics that discusses the collection, classification, and interpretation of discrete data. The primary purpose of using statistics when reporting laboratory data is to determine the average value and repeatability of the experimental results which is indicated by standard deviation. The minimum number of repetitions needed to conduct statistical analysis depends on the margin of error, the confidence level, and other variables.

Generally, to generate results with smaller error margin and higher confidence level, the test should be repeated more. However, sometimes in this Lab three repetitions are considered enough for the sake of educational purposes. The statistical analysis helps determine the average values and degree of consistency from test to test.

1.8.2   DATA DESCRIPTION

Data may be described in a variety of ways. Perhaps the most efficient manner of presenting data includes the use of graphics, displaying the data as a point plot or histogram. In addition, the data itself is given characteristic numbers to describe the data distribution. Some common numbers given to data are as follows.

Arithmetic Mean

The arithmetic mean is the average of a set of data, and is calculated by:

$$\mu = \frac{\sum x}{n} \tag{1.1}$$

n = the number of data points
x = sum of collected data
μ = arithmetic mean

Median

The median is the middle value of a data set when the data set is sorted from smallest to highest value. In the case of odd number of data, the median is the average of the two middle values.

Range

The range is the difference between the highest and lowest values in a data set.

Variance (σ^2)

The variance is a measure of central tendency (a measure of the tendency of the data to group about a certain value), and is determined by:

$$\sigma^2 = \frac{\sum (x-\mu)^2}{n}. \tag{1.2}$$

When the entire set (or population) of data is used to calculate the variance.

Standard Deviation (σ)

The standard deviation is commonly the most-used measure of central tendency and is defined as the square root of the variance.

Simple Linear Regression

Linear regression is used when the relationship between variables must be determined. Linear regression suggests that two quantities are related, such that increasing (or decreasing) the x-component increases (or decreases) the y-component. In a simple linear regression model, we attempt to fit our data to a line described by:

$$y = mx + b. \tag{1.3}$$

To obtain a best fit model, we will use the least square method, which minimizes the Sum of the Squared Errors (SSE) to the best fit line. In this method, the best fit line is given by:

$$y = b_1 x + b_0, \tag{1.4}$$

where $b_1 = \dfrac{n\sum(x_i y_i) - \sum x_i \sum y_i}{n\sum(x_i^2) - (\sum x_i)^2}$ and $b_0 = \bar{y} - b_1 \bar{x}$. Also, $\bar{x}$ = the arithmetic mean of x values and $\bar{y}$ = the arithmetic mean of y values.

Coefficient of Correlation (r)

The Coefficient of Correlation is a measure of how well the data fits your model, with $r = 1$ being a perfect fit, and $r = 0$ being no correlation. The Coefficient of Correlation is calculated by:

$$r = \frac{n\sum(x_i y_i) - \sum x_i \sum y_i)}{\sqrt{[n\sum(x_i^2) - (\sum x_i)2][n\sum(y_i^2) - (\sum y_i)^2]}} \tag{1.5}$$

The coefficient of determination is defined as r^2.

1.9   UNCERTAINTY ANALYSIS OF EXPERIMENTAL RESULTS

In engineering the word "error", when used to describe an aspect of measurement, does not necessarily carry the connotation of mistake or blunder (although it can!). Error in a measurement means the inevitable uncertainty that attends all measurements. We cannot avoid errors in this sense. All we can do is to ensure that they are as small as required by codes, standards, etc. and that we have a reliable estimate of how small they are.

All experimental measurements are carried out using measuring tools and since every measuring tool has a limit to its accuracy, there is some uncertainty involved in any measurement. It is quite easy to determine the uncertainty when you use a tool to directly measure a value. For instance, if you use a ruler, that has 1/8″ marks, to measure the thickness of a sheet of paper, you may estimate that the paper is about 1/10 of the smallest increment of measurement of your ruler or 0.0125″. However, you cannot certainly say that this is an accurate measurement. Actually, the error margin in your measurement can never be less than half of the smallest increment or 0.0625″. Therefore, you should report the thickness of paper as 0.0125″ ± 0.0625″. The "± 0.0625" inches is the uncertainty of your measurement with the ruler.

Obviously, in this case the error is huge compared to the measured value which suggests that you need to use a more precise measuring device. Should you replace the ruler with a micrometer that reads values to ± 0.0001" you will be able to measure the thickness of a sheet of copy paper at 0.0042± 0.0001". In this case, the uncertainty is less than 2.5% of the measured value which is a more precise measurement.

Often, we have two or more measured quantities that are combined mathematically to generate a result. For instance, velocity is computed by dividing the length, measured by ruler, by time, measured by chronometer. Combining these uncertain values propagates the error and makes it necessary to measure the uncertainty of the final result such that it reflects the uncertainty of each measured quantity. **Uncertainty analysis** is the procedure used to determine the propagation of uncertainty when calculating a quantity like "velocity" from multiple measured quantities.

One common method of measuring uncertainty is proposed by Kline and McClintock. This method is based on a careful specification of the uncertainties in the various primary experimental measurements. If R = $f(x_1, x_2, ..., x_{n-1}, x_n)$ is the resultant quantity we want and $x_1, x_2, ..., x_{n-1}, x_n$ are the primary measured quantities with the uncertainties $w_1, w_2, ..., w_{n-1}, w_n$, the equation used for this method is given below.

$$W_R = \left[\left(\frac{\partial R}{\partial x_1} w_1\right)^2 + \left(\frac{\partial R}{\partial x_2} w_2\right)^2 + \cdots + \left(\frac{\partial R}{\partial xn} w_n\right)^2 \right]^{1/2}, \qquad (1.6)$$

where, W_R = the uncertainty in the experimental result,

   R = the given function of the independent variables $x^1, x_2, ..., x_n$,

   R = R $(x^1, x_2, ..., x_n)$, and

   $w^1, w_2, ..., w_n$ = the uncertainty in the independent variables.

Example of Uncertainty Calculation

In the first experiment, fluid properties, students need to determine settling velocity of three types of beads that were fallen into water. To use Stock's Law (Equation 2.1) and calculate viscosity of the fluid in a tube. Students recorded the time in which small acrylic beads had traveled 1.00-m length of a clear tubes ten trials.

In this example, determine mean, median, variance, standard deviation, and uncertainty of velocity.

Raw Data: Example of Statistical Analysis

Table 1.2: **Raw Data for Small Acrylic Beads Traveled Time in One Meter Length of a Tube**

Trial	Time (sec)	Velocity (m/sec)
1	8.89	0.11249
2	9.34	0.10707
3	9.47	0.10560
4	9.48	0.10549
5	9.56	0.10460
6	9.58	0.10438
7	9.68	0.10331
8	10.23	0.09775
9	10.55	0.09479
10	12.26	0.08157
Mean	**9.90**	**0.10170**
Median	**9.57**	**0.10449**
Std	**0.90**	**0.00813**

Statistical Analysis

The raw data collected from small acrylic bead test are sorted from smallest to largest. The velocity is calculated for each trial. It is worth noting that because the length is constant, sorting data based on time automatically sorts velocities. The mean time is computed to 9.90 sec:

$$Mean\ Time = \frac{8.89 + 9.34 + 9.47 + 9.48 + 9.56 + 9.58 + 9.68 + 10.23 + 10.55 + 12.26}{10} = 9.90\ sec.$$

Similarly, the average velocity is 0.10170 m/sec. Since the number of trials is odd, the median is the average of two middle numbers after sorting.

$$Mean\ Time = \frac{9.56 + 9.58}{2} = 9.57\ sec$$

$$Mean\ Velocity = \frac{0.10460 + 0.10438}{2} = 0.10449\ m/sec$$

$$Variance\ of\ Time = \sigma_{time}$$

$$\sigma_{time} = \frac{(8.89-9.90)^2 + (9.34-9.90)^2 + (9.47-9.90)^2 + (9.48-9.90)^2 + (9.56-9.90)^2 + (9.58-9.90)^2 + (9.68-9.90)^2 + (10.23-9.90)^2 + (10.55-9.90)^2 + (12.26-9.90)^2}{10}$$

$$\sigma_{time} = 0.81$$

Standard Deviation of Time $= \sigma = \sqrt{\sigma^2} = 0.90$ sec

Similarly, one can calculate the variance of velocity at 0.000066 and the standard deviation of velocity at 0.00813 m/sec.

Uncertainty Analysis

Velocity is measured from the formula

$$V = \frac{L}{t},$$

which includes two directly measured quantities, L and t which denote the length and time, respectively. Therefore, to calculate the uncertainty, it is necessary to calculate the partial derivative of velocity with respect to these two quantities:

$$\frac{\partial V}{\partial L} = \frac{1}{t},$$

$$\frac{\partial V}{\partial t} = -\frac{L}{\partial t^2}.$$

Usually, manufacturers report the uncertainty of any measuring tool in its data sheet. In the absence of such information, one can assume that the uncertainty of a measuring device equals one half of its smallest increment. For instance, in this test a metric ruler is used to measure the length. Given that the smallest mark on the ruler was 1 mm, the uncertainty of the length (W_L) equals 0.5 mm. It is worth mentioning that to assure the units' consistency, all quantities should be reported in standard SI units. Therefore,

$w_L = 0.0005$ m,

$w_t = 0.005$ sec.

Plugging in the aforementioned quantities in the Kline and McClintock equation results in:

$$W_V = \sqrt{\left(\frac{\partial V}{\partial L} \times w_L\right)^2 + \left(\frac{\partial V}{\partial t} \times w_t\right)^2} = \sqrt{\left(\frac{1}{t} \times 0.0005\right)^2 + \left(\frac{-1}{t^2} \times 0.005\right)^2}.$$

Table 1.3: Uncertainty Calculation for Velocity of Beads

Trial	Time (sec)	Velocity (m/sec)	WV (sec)
1	8.89	0.11249	8.47E-05
2	9.34	0.10707	7.84E-05
3	9.47	0.10560	7.68E-05
4	9.48	0.10549	7.67E-05
5	9.56	0.10460	7.57E-05
6	9.58	0.10438	7.54E-05
7	9.68	0.10331	7.43E-05
8	10.23	0.09775	6.83E-05
9	10.55	0.09479	6.53E-05
10	12.26	0.08157	5.26E-05
Mean	9.90	0.10170	7.28E-05
Median	9.57	0.10449	7.56E-05
Std	0.90	0.00813	8.39E-06

The uncertainty of measuring velocity depends on two quantities, length and time, and varies from trial to trial. The average uncertainty of velocity is 7.28 E-5 (sec) and hence the average velocity should be reported as follows:

$$V_{avg} = 0.10170 \pm 7.28E\text{-}5 \text{ (m/sec).}$$

This can be rounded and rewritten in the next format.

$$V_{avg} = 0.10170 \pm 0.00007 \text{ (m/sec).}$$

1.9.1   ERROR ANALYSIS

$$Error = (Experimental\ value) - (Theoretical\ value) \tag{1.7}$$

$$Percent\ Error = \frac{|Error|}{Theoretical\ value} * 100 \tag{1.8}$$

The expected results are found from tabulated data from a reliable source or determined by theoretical equations (i.e., Cimbala and Cengel, 2018).

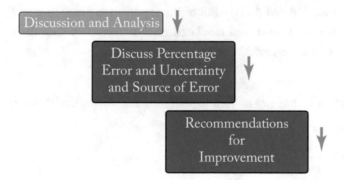

1.10   CONCLUSION

Every report must have a conclusion as a summary of your work and major findings. If your aim were to verify a law, state whether you have verified the law or not. If the aim were to measure a particular quantity (i.e. fluid viscosity), state the final measured value of the quantity in the conclusion. The conclusion is a summary of what you did, what you determined, and what you thought about the results. Try to explain any difference between your result and the initial expectations of the experiment as well as theoretical values.

1.11   REFERENCE

List any references where you found information. The textbook and lab manual are often listed as a reference for experiment pictures and equations.

1.12   APPENDICES

The appendices include some supplementary information, i.e., the raw data, sample calculation, some extra graphs, or tables from results, etc.

1.12.1 LAB REPORT CHECK LIST

- ✓ Cover Page
- ✓ Abstract: less than 250 words including summary of objective(s), procedure, quantified results.
- ✓ Table of Contents/Figures/Tables
- ✓ Introduction: provides background and explains why the experiment was performed.

✓ Procedure: detailed description of the experiment procedure including apparatus and equipment used, and tabulated measurements so that the test can be conducted by the reader(s).

✓ Results

✓ Discussion: summary of your experiment and results with your recommendation.

✓ References

✓ Appendices

1.13   HOW TO WRITE A LABORATORY MEMO

Sometimes a long technical laboratory report is not needed because of performing a short experiment or conducting a simple test. In engineering, a laboratory memorandum, or simply a lab memo, is a brief version of the technical laboratory report. The main goal of lab memo is to describe the objective, procedure, and results of an experiment to test the accuracy of a hypothesis or to answer a question. Often a lab memo length is up to one-page equivalent of text and appendices includes graphs, figures, sample calculation, etc. Students usually find writing a lab memo more difficult than a technical report because they need to summarize all the experiment elements. The structures of lab memo are as follows.

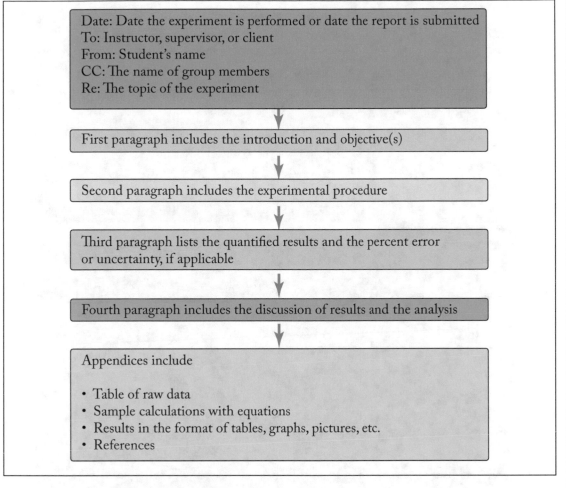

Date: Date the experiment is performed or date the report is submitted
To: Instructor, supervisor, or client
From: Student's name
CC: The name of group members
Re: The topic of the experiment

First paragraph includes the introduction and objective(s)

Second paragraph includes the experimental procedure

Third paragraph lists the quantified results and the percent error or uncertainty, if applicable

Fourth paragraph includes the discussion of results and the analysis

Appendices include

- Table of raw data
- Sample calculations with equations
- Results in the format of tables, graphs, pictures, etc.
- References

Example of a Good Lab Memo—Venturi Lab

Date: October 29, 2019
To: Dr. Robabeh Jazaei
From: Rebecca M.
CC: Alex E., and Kaitlyn G.
Re: The Energy Grade Line and Hydraulic Grade Line

One of the ways to visually observe the major or minor losses within a piping system is to draw the energy grade line (EGL) and hydraulic grade line (HGL). The purpose of this lab was to apply Bernoulli's equation and observe the visual EGL and HGL changes in various cross sections using a venturi apparatus.

To begin, the venturi apparatus was positioned on the hydraulic bench and leveled using the leveling screws. The pump was started and adjusted to provide a desired flow. The corresponding flow was measured using a stopwatch and the piezometer of the hydraulic bench. The venturi piezometer values were recorded from columns A–L for the corresponding flow. The hydraulic bench was closed and allowed to drain after each trial. This process was repeated for a total of three trials.

The Energy Grade Line (EGL) and Hydraulic Grade Line (HGL) was determined for three initial piezometer elevations at location A, 300, 250, and 150 mm. The calculated EGL and HGL values at each location shown in Figure 1 can be found in Table A.1 in Appendix A. Sample calculations for each of these values are listed in Appendix B. The relationship between the HGL/EGL and distance from the datum can be seen in Figure 2.

Bernoulli's equation was used to express the flow rate based on the change in pressure and area between inlet A and the throat located at point D. This equation can be found in Appendix B along with a sample calculation for Trial 1. The percent error between the calculated flow and measured flow was 4.85%.

The uncertainty in the flow rate can be accounted for by the precision of the instrument and potential air trapped in the apparatus. Since the piezometer is read in liters and there are not smaller units on the device, each trial may produce the same volume, but is inaccurately read because of the low precision of the instrument. To improve the precision, the apparatus could be fitted with increments in milliliters. Air trapped in the apparatus could result in an inaccurate reading of the heights on the piezometer. If the air is not noticeably visible, the trial will produce inaccurate results. To adjust for this error, the valve at the top of the apparatus can be opened slightly to release pressure in the tubes caused by the air. This also corrects the height of the water, producing more accurate results.

Color code:
Introduction and objectives
Procedure
Results
Percentage error and uncertainty
Discussion and recommendation

CHAPTER 2

Fluid Properties

EXPERIMENT: FLUID PROPERTIES

2.1 INTRODUCTION

Fluid properties dominantly depend upon the viscosity of the fluid. Viscosity is defined as the internal resistance of a fluid to motion and is expressed as the ratio of shear stress to the rate of deformation. The absolute viscosity can be solved for using a form of Stokes' Law which is proportional to the density of the fluid at a given temperature, the radius of the spherical object, and the velocity of that object through the fluid. For this experiment, the viscosity of water will be determined using various sizes of acrylic, plastic, and steel beads through a falling bead test. This experiment can be performed with other materials as well.

2.1.1 REQUIRED EQUATIONS

Stokes' Law

$$\mu = \frac{2\Delta\rho g r^2}{9V} \tag{2.1}$$

μ = Viscosity ($\frac{kg}{m \cdot sec}$)

$\Delta\rho$ = Difference in density between object and fluid ($\frac{kg}{m^3}$)

g = Acceleration due to gravity ($\frac{m}{sec^2}$)

r = Radius of object (m)

v = Settling velocity ($\frac{m}{sec}$)

2.2 OBJECTIVE

The purpose of this experiment is to further understand viscosity of various liquid (i.e., water, oil, etc.) by developing an experimental procedure that uses Stokes' Law.

2.3   EQUIPMENT

1. Tubes

2. Spheres (three size of steel and acrylic bead)

3. Stopwatches

4. Thermometer

5. Ruler

6. Caliper

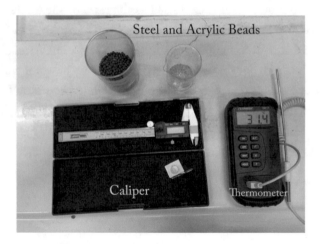

Figure 2.1: Caliper, thermometer, and steel/acrylic bead.

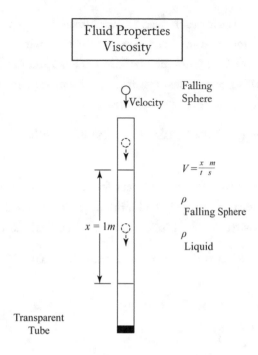

Figure 2.2: Fluid properties lab set-up.

2.4   PROCEDURE

1. Start by placing two pieces of tape on the rubber tube about 1 m apart. Measure the actual distance between those pieces of tape and record the value (Figure 2.2).

2. Determine an average weight per bead by using 10 beads (acrylic, steel, and plastic) and dividing the value by 10.

3. Use a caliper to measure the average radius of each bead and record the value (Figure 2.1).

4. Measure the water temperature with a thermometer (°C) in Figure 2.1.

5. Fill the tube with water until the water level is a little above the top piece of tape (Figure 2.2). Measure the temperature to determine the corresponding density from the textbook table (Cimbala and Cengel, 2018, Appendix 1, Table A-3).

6. Record the time it takes for each bead to travel from the top piece of tape to the bottom piece. Assign one team member to record the time with a stopwatch as the bead is falling (Figure 2.2). One team member drops a bead and the third records the data. Start with performing one or two test trials of the procedure to find ways to ensure consistency and accuracy in your procedure.

7. Perform a suitable number of trials (10 drops) for each bead type using the experimental procedure.

8. Enter your information into a spreadsheet for an analysis. Be sure to include all the information necessary for analysis, including the theoretical density of the fluid, bead diameter, bead density, etc. Use proper units in your calculation.

9. Compare the theoretical and experimental values by calculating the percent error.

2.4.1 EXPERIMENTAL DATA

Table 2.1: Record Time for Each Bead

	Time (sec)		
Trial	Acrylic (small)	Acrylic (large)	Steel
1			
2			
3			
4			
5			
6			
7			
8			
9			
10			
Average			

Table 2.2: Record Object and Fluid Properties

Bead Type	Bead Mass	Bead Diameter	Bead Radius	Bead Density	Fluid Density	Delta Density	Velocity	Viscosity
Unit	g	mm	m	kg/m^3	kg/m^3	kg/m^3	m/sec	kg/m.sec
Steel								
Acrylic (small)								
Acrylic (large)								

2.5   LAB ASSIGNMENT

Use the template provided for a basic mamo. Your report should include the following.

Find the theoretical value for water viscosity from the textbook and compare with the experimental results.

- Provide sample calculation for a single trial.

- Calculate percentage error.

- Determine uncertainty.

- Describe why the experimental and theoretical value is different.

- Write your recommendation to achieve better results in the discussion paragraph.

- Include any graphs, tables, and raw data in the appendices.

CHAPTER 3

Fluid Statics—Pressure Measurement and Hydrostatics

3.1 EXPERIMENT: PRESSURE MEASUREMENT WITH MANOMETERS AND PIEZOMETERS

3.1.1 INTRODUCTION

Pressure is defined as the normal force exerted by a liquid or gas per unit area. In solids, pressure is referred to as normal stress. There are different methods to measure various pressures including atmospheric, absolute and gage. The pressure measurement devices that are widely used include:

1. the barometer, which measures the atmospheric pressure (Figure 3.1);

2. piezometer or pressure tube which measures the static pressure head (Figure 3.1); and

3. manometer, also known as a U-tube, which measures the pressure in one end or the differential pressure between two points:

   a. simple manometer (Figure 3.2); and

   b. differential manometer (Figure 3.2).

A simple manometer apparatus will be used to gather information on the heights of the known liquids to determine the experimental pressure of the system.

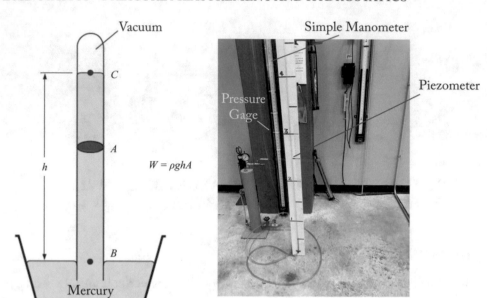

Figure 3.1: Barometer (left-side picture) and piezometer (right-side picture).

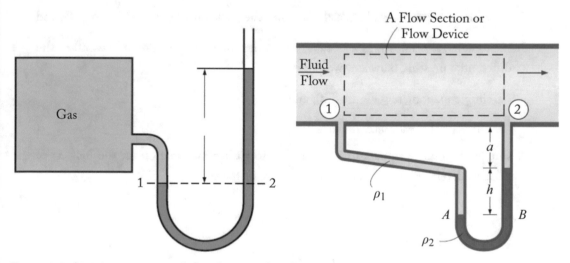

Figure 3.2: Simple manometer (left-side picture) and differential manometer (right-side picture).

Required Equations

Before you start the experiment, think about how to develop the equations for the pressure in the piezometer and differential manometer. The sketch for the piezometer and simple manometer for the test set-up is shown in Figure 3.3. To determine the theoretical pressure in the system based on the measured heights, look at Example 3-3 in the textbook, and write the equations. Think about when the pressure increases or decreases. Find the theoretical value for gage pressure. At the end of the experiment, you will compare the theoretical and experimental values for the pressure.

Piezometer

$$P = \rho g h + P_{atm} \tag{3.1}$$

Simple Manometer

$$P_{atm} + \rho_1 g h_1 + \rho_2 g h_2 + \rho_3 g h_3 = P \tag{3.2}$$

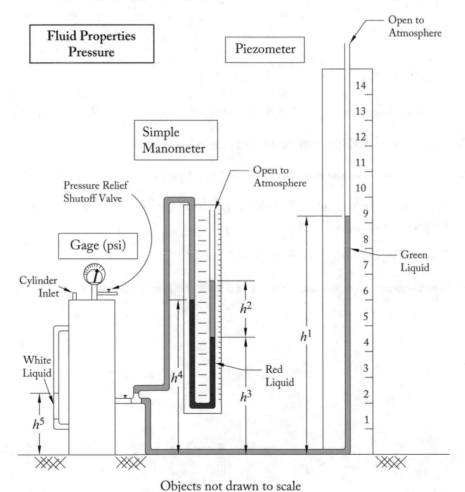

Objects not drawn to scale

Figure 3.3: Simple manometer and piezometer set-up.

3.1.2   OBJECTIVE

The purpose of this lab is to further understand the use of a simple manometer and how pressure differences are measured through the change in elevation of a fluid with a known density.

3.1.3   EQUIPMENT

1. Simple manometer (Figure 3.2)

2. Pressurized cylinder

3. Air pump

4. Tape measure

3.1.4   PROCEDURE

1. Pump air from a pressurized cylinder, which is connected to a simple manometer

2. Make sure enough pressure build up in the cylinder.

3. Use the same pressure gage for all trials (e.g., 4 psi) and record the heights h_1–h_5.

4. Measure all the heights h_1–h_5 (Figure 3.4) from the floor (h_5 is the empty height in the pressure cylinder).

5. Record the pressure gage in psi (Figure 3.5).

6. Open the air valve to release the air and start the next trial from step 1.

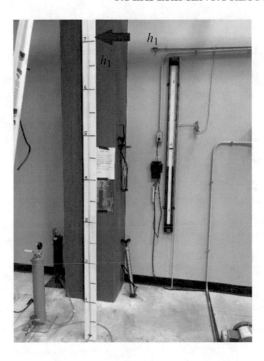

Figure 3.4: Measure h1 from the floor.

Figure 3.5: Pressure cylinder, pressure gage, height from the floor.

3.1.5   EXPERIMENTAL DATA

Table 3.1: Experimental Data for Manometer and Piezometer

	h_1	h_1	h_1	h_1	Pressure Gage
Unit	ft	ft	ft	ft	psi
Trial 1					
Trial 2					
Trial 3					
Trial 4					
Trial 5					
Trial 6					
Trial 7					
Trial 8					
Trial 9					
Trial 10					

Table 3.2: Experiment Constants Data

Constant	Value	Unit
Density of water	1.94	Slug/ft^3
S.G. Green Liquid	1.14	-
S.G. Red Liquid	2.95	-
*Note: The units for calculation in this lab is in BGS not SI.		

3.1.6   ASSIGNMENT

1. Use the template provided for basic lab memo for this experiment.

2. Assume atmospheric pressure is negligible. Write proper equations to determine the theoretical value of the pressure—(Equations 3.1 and 3.2)—one for the piezometer and one for manometer, using variables from Tables 3.1 and 3.2.

3. Conduct ten trials of the experiment and compare the pressure gage from both equations.

4. Calculate the percentage error and graph it as a function of pressure.

5. Determine uncertinity

6. Which device has the minimum percentage error with the gage reading.

7. If the green liquid is changed to kerosene and the red liquid is changed to glycerin, how much the pressure will be changed by percentages?

3.2   EXPERIMENT: HYDROSTATIC FORCES ON A PLANAR SURFACE

3.2.1   INTRODUCTION

Hydrostatic force is crucial in many aspects of civil engineering such as dams, storage tanks, ships, and water towers as a hydrostatic force is present when any object is submerged in a fluid. Hydrostatic forces represent the force perpendicular to the surface of a body submerged in a fluid. The principles of hydrostatic forces are relevant in many aspects of civil engineering such as the design of dams, storage tanks, and ships. The pressure forces on an object can be combined into one resultant force (F_R) which acts at the center of pressure (y_p) of an object. The hydrostatic pressure on an object increases in proportion to depth as expressed in the equations below.

Required Equations

Pressure

$$P = P_o + (\rho g h) \tag{3.3}$$

P_o = Atmospheric pressure (Pa)

ρ = Density of water ($\frac{kg}{m^3}$)

g = Acceleration due to gravity ($\frac{m}{sec^2}$)

h = Vertical distance from the fluid surface to the object's centroid (m)

Hydrostatic Force

$$F_R = (\rho g h_c)A \tag{3.4}$$

ρ = Density of water ($\frac{kg}{m^3}$)

g = Acceleration due to gravity ($\frac{m}{sec^2}$)

h_c = Vertical distance from the fluid surface to the object's centroid (m)

A = Area of submerged face (m^2)

Center of Pressure

$$y_p = y_c + \frac{I_{xx,c}}{y_c A}$$

(3.5)

y_c = y-coordinate of the centroid (m)

$I_{xx,c}$ = Second moment of area about x-axis passing through the centroid (m³)

A=Area (m²)

Centroidal Moment of Inertia for a Rectangle

$$I_{xx,c} = \frac{bh^3}{12}$$

(3.6)

3.2.2 OBJECTIVE

The purpose of this lab is to determine the hydrostatic force and center of pressure on a planar surface at varying angles (Figure 3.6).

3.2.3 EQUIPMENT

Constant measurements of the apparatus in Figure 3.8

$L = 28$ cm, distance from weight hanger to pivot

$b = 7.6$ cm, width of vertical quadrant face (planner surface)

$d = 10$ cm, height of vertical quadrant face

$h = a + d = 20$ cm

y = depth of water in each trail

m = mass on scale (g)

$g = 9.81$ (m/sec²)

Assume density of water is 1,000 kg/m³ or measure the fluid temperature with thermometer then find the associated density.

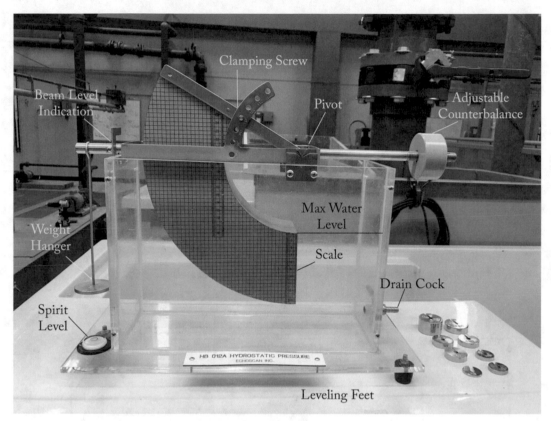

Figure 3. 6: Hydrostatic pressure apparatus.

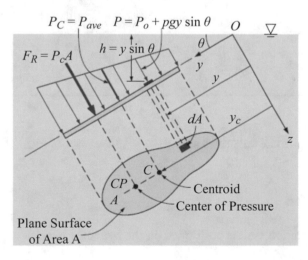

Figure 3.7: Centroid and height for pressure measurement on an inclined plane surface completely submerged in water.

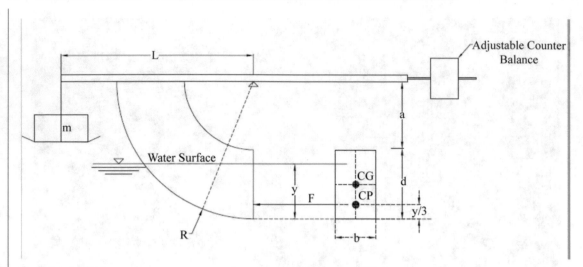

Figure 3.8: Hydrostatic pressure on a vertical surface (Ahmari and Kabir, 2019).

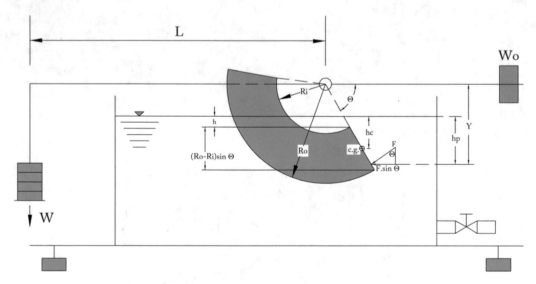

Figure 3.9: Hydrostatic pressure on a tilted surface.

3.2.4   PROCEDURE

1. The empty flotation tank is positioned on the hydraulic bench.

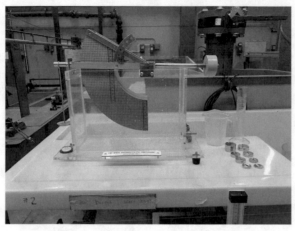

Figure 3.10: Hydrostatics apparatus.

2. Spirit level is accurately leveled by adjusting the leveling feet until the built-in circular spirit level indicates that the water tank is leveled in both planes.

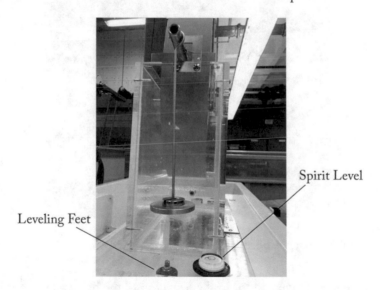

Figure 3.11: Hydrostatics apparatus required adjustment with spirit level.

3. The counter-balance weight is then moved until the balance arm is horizontal with the beam level indicator.

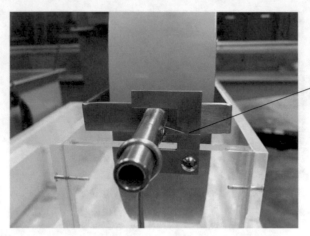

Figure 3.12: Level indicator.

4. Water is added to the tank until reaches 1 cm depth (y).

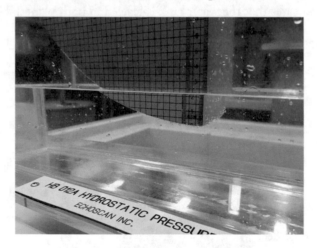

Figure 3.13: Hydrostatics apparatus filled with water at specific depth.

5. Appropriate mass is then added to the hanger until the balancing arm is leveled.

Equivalent
Mass

Figure 3.14: Required mass and apply as a counterbalance with hydrostatics force.

6. The height and mass are recorded in Table 3.4.

7. Perform an experiment using the laboratory apparatus (ten trials each) to determine the moment about pivot for the planar surface ($\theta = 90$).

8. Repeat steps 1–5 for the angled planner surface ($\theta = 70$). Perform an experiment using the laboratory apparatus (five trials each) to determine the moment about pivot for an angled planar surface.

9. Release the water valve, remove the weights, and clean up any spilled water at the end of experiment.

10. Make sure you put the apparatus and other equipment back to the allocated area.

3.2.5   EXPERIMENTAL DATA

Table 3.3: Experimental Data for 90°

Depth of Water	Trial 1	Trial 2	Trial 3	Trial 4	Trial 5	Trial 6	Trial 7	Trial 8	Trial 9	Trial 10	Average Mass on Hanger (g)
1											
2											
3											
4											
5											
6											
7											
8											
9											
10											

Table 3.4: Experimental Data for $\theta = 70°$

Depth of Water	Trial 1	Trial 2	Trial 3	Average Mass on Hanger (g)
1				
2				
3				
4				
5				

3.2.6   LAB ASSIGNMENT

Use the template provided for technical lab report for this experiment. Your report should include the following:

- table(s) of raw data;

- table(s) of results including theoretical and experimental hydrostatic force as well as percentage error;

- plot depth of immersion (x-axis) vs. hydrostatics force (y-axis) for both angels with the error bar;

- plot depth of immersion (x-axis) vs. mass (y-axis);

- calculate theoretical hydrostatics force acting on planar surfaces for $\theta=90°$ and $70°$ by writing two equations with proper notation (h_c, y_c, γ, b, and θ);

- calculate theoretical center of pressure by writing an equation for the moment of the hydrostatic force about point O as a function of (h_c, y_c, γ, b, and θ);

- provide sample calculation for both angles;

- discuss the effect of angle on hydrostatics force; and

- discuss the variations of hydrostatic force with depth of immersion.

3.3 EXPERIMENT: ARCHIMEDES' PRINCIPLE AND BUOYANT FORCE

3.3.1 INTRODUCTION

You might feel lighter when you are swimming than when you are on the ground. This is because upward pressure exerted from water. This is true for other objects when they are in a liquid compared to being in the air as well. This pressure is called the Buoyant force.

3.3.2 OBJECTIVE

The aim of this experiment is to determine the weight of steel and aluminum cylinders in water and air. Also, to determine the density by utilizing Archimedes' Principle, a physical law of buoyancy, that demonstrates the equivalent magnitude of the buoyant force. The volume of displaced fluid is equivalent to the volume of an object submerged in a fluid. The objective of this experiment is to apply Archimedes' Principle to an aluminum and steel cylinder.

Required Equations

Theoretical Buoyant Force

$$F_b = \rho_{fluid} V_{displaced},$$

(3.7)

where

F_b = buoyant force (N)

ρ_{fluid} = density of fluid (kg/m^3)

g = gravity (m/sec^2)

$V_{displaced}$ = volume of fluid displaced (m^3)

Experimental Buoyant Force

$$F_b = (m_{actual} - m_{apparent})g, \tag{3.8}$$

where

F_b = buoyant force (N)

m_{actual} = mass of object in air (kg)

$m_{apparent}$ = mass of object in water (kg)

g = gravity (m/sec^2)

3.3.3   PROCEDURE

1. Measure the temperature of water by using a thermometer then find the corresponding water density.

2. Use spring scale to measure the weight of steel and aluminum cylinder in water and air, as shown in Figure 3.14 (ρ_{steel} = 8,050 kg/m^3, $\rho_{Aluminum}$ = 2,700 kg/m^3).

3. A graduated cylinder with a volume of known water is used to calculate the volume of displacement to further determine the theoretical buoyant force on each cylinder and theoretical density.

3.3.4   EQUIPMENT

1. Spring scale

2. Graduate cylinder

3. Two steel and aluminum cylinders

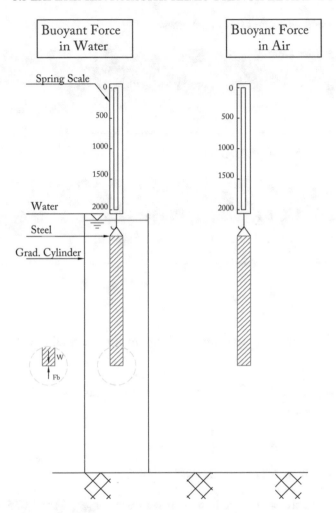

Figure 3.15: Buoyancy force apparatus sketch.

3.3.5   EXPERIMENTAL DATA

Table 3.5: Mass and Volume Measurement

	Mass in Air (g)	Mass in Water (g)	Volume Displaced (mL)
Aluminum			
Steel			

Table 3.6: Theoretical and Measured Buoyant Forces

	Theoretical Buoyant Force (N)	Experimental Buoyant Force (N)	Percent Error
Aluminum			
Steel			

Table 3.7: Theoretical and Measured Densities

	Theoretical Density (kg/m³)	Experimental Density (kg/m³)	Percent Error
Aluminum			
Steel			

3.3.6   LAB ASSIGNMENT

1. Compare theoretical and experimental buoyant force as well as density for steel and aluminum.

2. Calculate the percentage error for buoyant force and discuss if density of objects affects the buoyant force.

3. Determine uncertainty.

4. What is the percentage of weight loss in water?

5. How much will the weight be lighter in sea water ($\rho_{sea\ water}$ = 1,025 kg/m³) by percentage?

6. Determine the force in the rope for both aluminum and steel cylinders.

CHAPTER 4

Fluid Dynamics

4.1 EXPERIMENT: IMPACT OF JET-LINEAR MOMENTUM APPLICATION

4.1.1 INTRODUCTION

Linear momentum is described as the quantity of motion of an object and is measured as the product of the objects' mass and velocity. This lab demonstrates the impact of a jet apparatus on different targets and the measured net force acting on each target. This allows us to further understand how hydraulic machinery, such as the Pelton Wheel Turbine, generates electricity. When high water pressure flows through a nozzle and hits the vane, pressure energy is converted to kinetic energy that falls into Newton's second law as expressed in the equations below.

Required Equations

Linear Momentum

$$F = ma = \dot{m}(v_1 - v_2) \tag{4.1}$$

$\dot{m}$ = Mass flow rate $\left(\frac{\text{kg}}{\text{sec}}\right)$

$v_1 - v_2$ = Change in velocity $\left(\frac{\text{m}}{\text{sec}}\right)$

Continuity Equation and Energy Equation

The combination of the energy and continuity equations is written as the flow rate entering a control volume equal to the flow rate exiting the control volume:

$$\text{For steady flow: } \sum \dot{m}_{in} = \sum \dot{m}_{out} \tag{4.2}$$

$$\text{For single steady-flow system: } \dot{m}_1 = \dot{m}_2 => \rho_1 V_1 A_1 = \rho_2 V_2 A_2 \text{ and } \rho_1 Q_1 = \rho_2 Q_2 \tag{4.3}$$

Combined Linear Momentum Equation

$$F = \rho Q \, (v_2 - v_1) \tag{4.4}$$

$$\rho = \text{Density } (\tfrac{\text{kg}}{\text{m}^3})$$

$$Q = \text{Flow rate } (\tfrac{\text{m}^3}{\text{sec}})$$

$$v_1 - v_2 = \text{Change in velocity } (\tfrac{\text{m}}{\text{sec}})$$

Impact of Jet in the y Direction (Fy)

$$F_y = \rho Q V (\cos \theta - 1) \tag{4.5}$$

$$F_y = \rho A V^2 (\cos \theta - 1) \tag{4.6}$$

$$\rho = \text{Density } (\tfrac{\text{kg}}{\text{m}^3})$$

$$Q = \text{Flow rate } (\tfrac{\text{m}^3}{\text{sec}})$$

$$V = \text{Velocity } (\tfrac{\text{m}}{\text{sec}})$$

$$\theta = 180° - \alpha$$

$\alpha = $ Target plate angle (flat target plate: $\alpha = 90°$, hemispherical target plate: $\alpha = 180°$, conical target plate: $\alpha = 60°$)

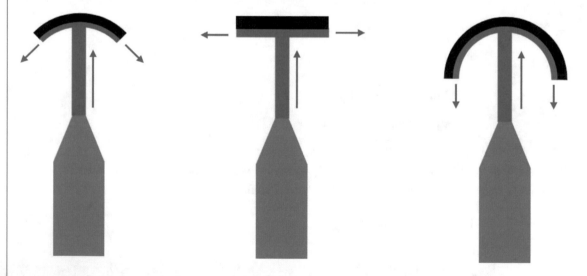

Figure 4.1: Free body diagram for control volume with 3 target plate angles (flat target plate: $\alpha = 60°$, hemispherical target plate: $\alpha = 90°$, conical target plate: $\alpha = 90°$).

From Equilibrium, keep in mind $F_y = W = mg$ where m is mass and g is gravity.

4.1.2   OBJECTIVE

The purpose of this lab is to determine the net force a jet apparatus exerts on three different styles of plates: flat, hemispherical, and conical.

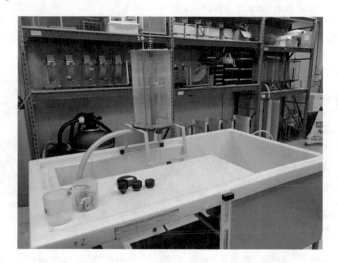

Figure 4.2: Impact of a jet apparatus.

Figure 4.3: Impact of a jet apparatus in action.

4.1.3   EQUIPMENT

1. Hydraulic Bench

2. Impact of a Jet Apparatus

3. A flat, hemispherical, and a conical target plate

4. Set of calibrated weights (50, 100, 200, 300, and 400 g)

(a)	(b)	(c)

Figure 4.4: (a) Pointer, weight carrier, and spirit level; (b) flat, hemispherical and 120° vane; and (c) replace a new vane.

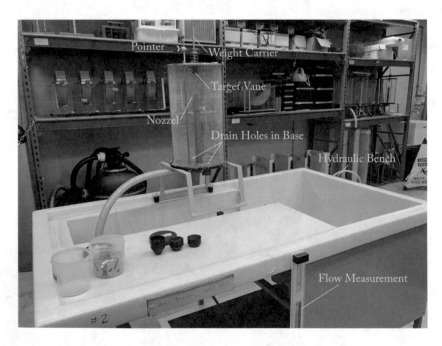

Figure 4.5: Impact of a jet apparatus details.

4.1.4   PROCEDURE

1. Position the apparatus on the hydraulic bench and level it with the level-sprit on the top of the apparatus (Figure 4.3).

2. Align the pointer with the weight carrier on the top of apparatus before applying any force from the jet of water (Figure 4.4a).

3. Choose the flat target plate ($\alpha = 90°$). See Figure 4.4b and 4.4c.

4. After replacing the proper target plate on the apparatus, ensure the top screws are secured and the pointer is still aligned with weight carrier (Figure 4.4a).

5. Start the pump to provide desirable, steady flow by opening the adjustable valve on the hydraulic bench.

6. A jet in the cylinder shoots water onto the surface of a plate at the top of the nozzle. Water runs down the side of the cylinder into a channel which feeds a drain and piezometer.

7. Add the appropriate masses to the weight carrier until it is realigned with the pointer.

8. Measure the corresponding flow. Use a stopwatch and record the volume of water in the piezometer of the hydraulic bench. (Note: $Q = \frac{V}{t}$ (L/sec))

9. Record the flow rate and mass.

10. Close the hydraulic bench valve and turn off the pump. Allow the apparatus to drain between trials.

11. Repeat the procedure with the hemispherical and conical target plates (Figure 4.4b).

4.1.5   EXPERIMENTAL DATA

Table 4.1: Experimental Data for Flat $\theta = 90°$

Flat $\theta = 90°$	Weight (g)	Volume of Water (L)	Time (sec)
Trial 1	50		
Trial 2	100		
Trial 3	200		
Trial 4	300		
Trial 5	400		

Table 4.2: Experimental Data for Hemispherical θ = 180°

Flat θ = 180°	Weight (g)	Volume of Water (L)	Time (sec)
Trial 1	50		
Trial 2	100		
Trial 3	200		
Trial 4	300		
Trial 5	400		

Table 4.3: Experimental Data for Conical θ=120°

Flat θ = 120°	Weight (g)	Volume of Water (L)	Time (sec)
Trial 1	50		
Trial 2	100		
Trial 3	200		
Trial 4	300		
Trial 5	400		

Table 4.4: Experiment Constants

Constant	Value	Unit
Density of Water	1000	kg/m^3
Nozzle Diameter	5	mm
* Note: Use SI units for this experiment.		

4.1.6   LAB ASSIGNMENT

- Draw free body diagram of three plates with corresponding theoretical forces and velocities.

- Plot a graph which velocity squared is in X-axis and applied weight is in Y-axis then determine the trend line of graph to find the slope.

- Compare the graph slope with theoretical slope (Equation 4.7).

$$S = \rho A (\cos \alpha + 1) \tag{4.7}$$

S = Slope of graph W vs. V^2

ρ = Density $(\frac{kg}{m^3})$

Q = Flow rate ($\frac{m^3}{sec}$)

V = velocity ($\frac{m}{sec}$)

α = Target plate angle

- Determine percentage error and standard error then explain why standard error is different for three target plates.

- Determine uncertainty.

- Plot applied weight (W) vs. the impact of jet (F_y)

- Explain how this experiment can demonstrate the conservation of momentum

- Why are the experimental and theoretical values different? What is the significant contributor in high standard error for one of the target plates?

4.2   EXPERIMENT: THE EGL AND HGL-VENTURI

4.2.1   INTRODUCTION

Pressure and velocity of a fluid vary as the fluid passes through different cross sections. Bernoulli's equation expresses that the summation of energy, including pressure head, velocity head, and elevation head remain constant along a streamline. A Venturi Meter (TecQuipment Ltd 2000) is used to visually demonstrate changes in the EGL and HGL through various cross-sections of the apparatus as well as energy loss (Figure 4.6). This device has been used to measure the pipe discharge (flow rate) throughout a contraction section for a long time. The flow remains constant; however, the velocity varies through each section and can further be calculated from the continuity equation.

Required Equations

Continuity Equation

$$Q = VA \qquad\qquad (4.8)$$

Q = Flow rate ($\frac{m^3}{sec}$)

V = Velocity ($\frac{m}{sec}$)

A = Cross-sectional area (m^2)

Bernoulli's Equation

$$\frac{P}{\gamma} + \frac{V^2}{2g} + Z = H = \text{Constant} \tag{4.9}$$

Table 4.5: EGL and HGL Terms

Terms	Head
$\frac{P}{\gamma}$	Pressure Head
$\frac{V^2}{2g}$	Velocity Head
Z	Elevation Head
H	Total Head
$\frac{P}{\gamma} + \frac{V^2}{2g} + Z$	Energy Grade Line (EGL)
$\frac{P}{\gamma} + Z$	Hydraulic Grade Line (EGL)

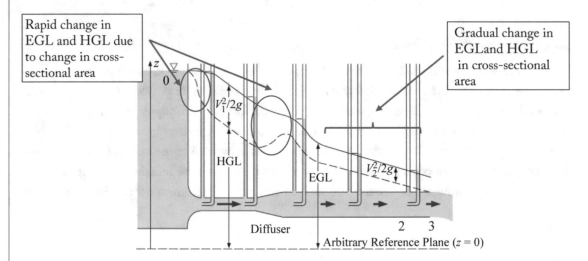

Rapid change in EGL and HGL due to change in cross-sectional area

Gradual change in EGL and HGL in cross-sectional area

Figure 4.6: EGL and HGL due to change in cross-sectional area and energy loss in constant cross-sectional area.

4.2.2   OBJECTIVE

The objective of this lab is to apply Bernoulli's equation (Equation 4.9) and visualize the EGL and HGL changes in various cross sections using the Venturi Apparatus. The corresponding piezometer height indicates the effect of velocity and pressure changes through the apparatus.

Figure 4.7: Bench top venturi apparatus.

4.2.3   EQUIPMENT

1. Hydraulic Bench

2. Venturi Apparatus

3. Stopwatch

Figure 4.8: Bench top venturi apparatus detail.

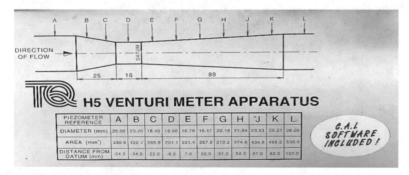

Figure 4.9: Venturi apparatus cross-sectional area from columns A–L (no "I").

4.2.4   PROCEDURE

1. Position the apparatus on the hydraulic bench and level it with the adjustable screws.

2. Tightly connect an adjustable hose to the hydraulic bench. Place a second hose in the hydraulic bench tub for water discharge, as shown in Figure 4.10.

3. Start the pump to provide desirable, steady flow by opening the adjustable valve on the hydraulic bench.

4. The piezometer tubes indicate the level of water from sections A–L in millimeters.

5. If there are some air bubbles in the piezometer tubes, a small air valve in manifold can be adjusted to have an accurate recording in the piezometer. Make sure the air valve is tightly closed after releasing trapped air from piezometer tubes.

6. Measure the corresponding flow. Use a stopwatch and record the volume of water in piezometer of the hydraulic bench in liter ($Q = \frac{V}{t}$, L/sec).

7. Record the piezometer heights from columns A–L and corresponding flow in Table 4.6. (Note: There is no "I" column.)

8. Close the hydraulic bench valve and switch off the pump. Allow the apparatus to drain after each trial.

9. Repeat the procedure with three different flow rates and record the piezometer height. Use the piezometer elevations in column A (300, 250, and 150 mm) and determine the experimental flow rate.

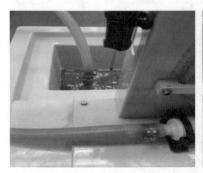

Figure 4.10: Venturi apparatus hose connection set-up.

4.2.5   EXPERIMENTAL DATA

Table 4.6: Experimental Data

Record Data	Unit (mm)	A	B	C	D	E	F	G	H	J	K	L	Flow (L/s)
	Diameter	26.00	23.20	18.40	16.00	16.79	18.47	20.16	21.84	23.53	25.24	26.00	–
	Area	530.9	422.7	265.9	201.1	221.4	267.9	319.2	374.6	434.8	499.2	530.9	-
	Distance	-54.0	-34.0	-22.0	-8.0	7.0	22.0	37.0	52.0	67.0	82.0	102.0	-
Trial 1	Elevation	300											
Trial 2	Elevation	250											
Trial 3	Elevation	150											

Table 4.7: Calculated Data for Column A = 300 mm

	A	B	C	D	E	F	G	H	J	K	L
$V = Q$ (m^3/sec)/A(m^2)											
P/γ (m)	0.3										
$V^2/2g$(m)											
Z (m)											
EGL: $P/\gamma + V^2/2g + Z$											
HGL: $P/\gamma + Z$											

Table 4.8: Calculated Data for Column A = 250 mm

	A	B	C	D	E	F	G	H	J	K	L
$V = Q$ (m^3/sec)/A(m^2)											
P/γ (m)	0.25										
$V^2/2g$(m)											
Z (m)											
EGL: $P/\gamma + V^2/2g + Z$											
HGL: $P/\gamma + Z$											

Table 4.9: Calculated Data for Column A = 150 mm

	A	B	C	D	E	F	G	H	J	K	L
$V = Q\,(m^3/s)/A(m^2)$											
P/γ (m)	0.15										
$V^2/2g$(m)											
Z (m)											
EGL: $P/\gamma + V^2/2g + Z$											
HGL: $P/\gamma + Z$											

*Note: Use SI units for this experiment.

4.2.6   LAB ASSIGNMENT

Note: Assume there is no head loss in the venturi apparatus.

1. Plot the energy grade line and hydraulic grade line. Compare the two graphs.

2. Derive an equation for the velocity based on the Bernoulli and continuity equations between points A and G.

3. Determine the theoretical velocity from the equation below for two points and compare to the experimental velocity for each column. Calculate the percent error for each.

$$V_2 = \frac{1}{\sqrt{1 - \frac{A_2^2}{A_2^1}}} * \sqrt{(2g\,(h_1 - h_2)} \tag{4.10}$$

V_2 = Velocity at point 1

A_1 = Cross-sectional area at point 1

A_2 = Cross-sectional area at point 2

h_1 = Column height at point 1

h_2 = Column height at point 2

4. Calculate the theoretical flow rate using the equation below and compare it to the experimental flow:

$$Q = \frac{A_2}{\sqrt{1 - \frac{A_2^2}{A_2^1}}} \sqrt{(2g\,(h_1 - h_2)} . \tag{4.11}$$

4.3   EXPERIMENT: FLOW FROM AN ORIFICE—DISCHARGE COEFFICIENT

4.3.1   INTRODUCTION

The coefficient of discharge or velocity can be used to measure pressure losses through an orifice. An orifice is commonly used in engineering applications to deliberately reduce flow rate and velocity. Vena contracta defines the point in a fluid streamline where all flow paths converge, and the area is at a minimum, as shown in Figure 4.11. Additionally, the characteristics of flow are changed throughout the orifice. This phenomenon affects the flow and velocity. The ratio of experimental flow/velocity to ideal flow/velocity are defined as the coefficient of discharge (C_d) and coefficient of velocity (C_v), respectively (Figure 4.12).

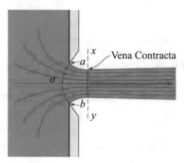

Figure 4.11: Vena contracta point in streamlines.

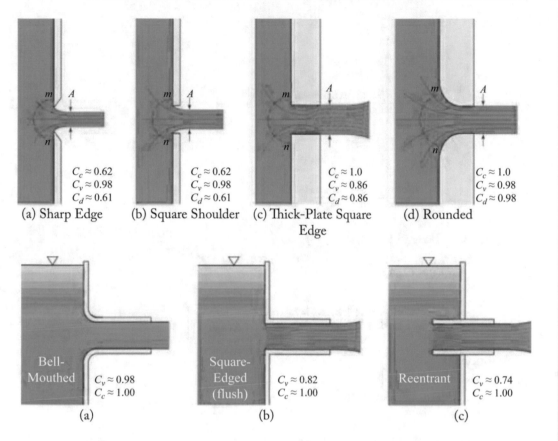

Figure 4.12: Example of coefficient of contraction, velocity, and discharge for different types of contraction.

Required Equations

Bernoulli's Equation for the Theoretical Velocity

$$\frac{P_1}{\gamma} + \frac{V_1^2}{2g} + z_1 = \frac{P_2}{\gamma} + \frac{V_2^2}{2g} + z_1 \qquad (4.12)$$

Note: Point 1 is located on the surface of the water in the orifice apparatus and point 2 is at the point of discharge.

Therefore, $\qquad 0 + 0 + h = 0 + \frac{V_2^2}{2g} + 0$

$$V_{\text{orifice}} = \sqrt{2gh}$$

Coefficient of Contraction, C_c

$$A_{jet} = C_c \; A_{orifice}$$

$$Cc = \frac{\text{Area of jet at vena contracta}}{\text{Actual area of orifice}} = \frac{V}{\sqrt{2gh}} \qquad (4.13)$$

Coefficient of Velocity, C_{vv}

$$V_{jet} = C_v \; V_{orifice}$$

$$C_v = \frac{\text{Velocity at vena contracta}}{\text{Theoretical velocity}} = \frac{V}{\sqrt{2gh}} \qquad (4.14)$$

Equation 4.4: Coefficient of Discharge, C_d

$$Q_{jet} = C_d \; Q_{orifice}$$

$$C_d = \frac{\text{Velocity Actual discharge in } t \text{ seconds}}{\text{Theoretical discharge in } t \text{ seconds}} = \frac{Q}{A\sqrt{2gh}} \qquad (4.15)$$

Note: Additional Equations

$$Q_{ideal} = A_{orifice} V_{ideal}$$

$$C_d = C_c \, C_v \qquad (4.16)$$

4.3.2   OBJECTIVE

The objective of this lab is to determine the discharge coefficient, C_d, and velocity coefficient, C_v, for three different orifices. The corresponding piezometer height indicates the effect of velocity and pressure changes through the orifice apparatus (Figure 4.13).

4.3.3   EQUIPMENT

1. Hydraulic bench

2. Orifice apparatus

3. Caliper

4. Three different orifices

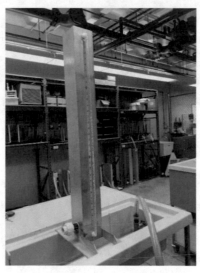

Figure 4.13: Bench top orifice apparatus.

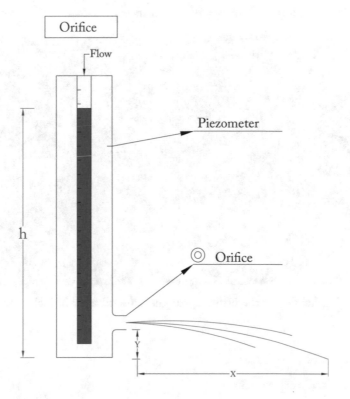

Figure 4.14: Orifice apparatus sketch.

4.3.4   PROCEDURE

1. Position the apparatus on the hydraulic bench and level it (Figure 4.15).

2. Measure the diameter of each orifice and record the values in Table 4.10.

3. Start the pump to provide steady, desirable flow by opening the adjustable valve on the hydraulic bench.

4. Allow the flow to reach a steady state.

5. Measure the corresponding flow. Use a stopwatch and record the volume of water from the piezometer on the hydraulic bench ($Q = \frac{V}{t}$, L/sec).

6. Record the height on the orifice apparatus, as shown in Figure 4.15, and the corresponding flow of the hydraulic bench (Table 4.11).

7. Close the hydraulic bench valve and switch off the pump.

8. Repeat the procedure with three different orifices.

Figure 4.15: Measuring heights on the orifice apparatus. (Note: Listed in inches on the apparatus.)

4.3.5   EXPERIMENTAL DATA

Note: Use SI units for this experiment.

Table 4.10: Orifice Diameters

D_1(mm)	D_2(mm)	D_3(mm)

Table 4.11: Experimental Data

	mm		L/sec
h_1		Q_1	
h_2		Q_2	
h_3		Q_3	
h_4		Q_4	
h_5		Q_5	
h_6		Q_6	
h_7		Q_7	
h_8		Q_8	
h_9		Q_9	
h_{10}		Q_{10}	

4.3.6   LAB ASSIGNMENT

1. Write Bernoulli's' equation and find the velocity at the orifice exit.

2. Measure and discuss the values for C_c, C_v, and C_d.

3. How does shape of the orifice affect the flowrate and coefficient of discharge?

CHAPTER 5

Internal Flow

5.1 EXPERIMENT: INTERNAL FLOW CLASSIFICATION— REYNOLDS NUMBER

5.1.1 INTRODUCTION

Reynolds number is a dimensionless number that British engineer, Osborne Reynolds (1842–1912), discovered over a century ago. The transition from laminar to turbulent flow depends on the geometry, surface roughness, flow velocity, surface temperature, and type of fluid, among other things. After exhaustive experiments in the 1880s, Osborne Reynolds discovered that the flow regime depends mainly on the ratio of inertial forces to viscous forces in the fluid. This ratio is called the Reynolds number and is expressed for internal flow in a circular pipe as follows (Cimbala and Cengel, 2018).

Required Equations

Reynold's Number

$$R_e = \frac{\text{inertial forces}}{\text{viscous forces}} = \frac{DV_{ave}}{v} = \frac{V_{ave}\rho D}{\mu} \tag{5.1}$$

V = Average flow velocity (m/sec)

D = Diameter (m)

ρ = Density $(\frac{\text{kg}}{\text{m}^3})$

$v = \frac{\mu}{\rho}$ = Kinematic viscosity $(\frac{\text{m}^2}{\text{sec}})$

μ = Absolute viscosity $(\frac{\text{m}^2}{\text{sec}})$

For the circular (Figure 5.1), pipe the flow regime is determined as follows:

    Laminar flow: Re < 2300

    Transitional flow: 2300 < Re < 4000

    Turbulent flow: Re > 4000

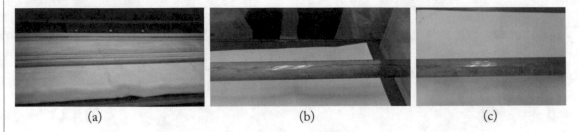

Figure 5.1: Reynolds number flow type: (a) laminar, (b) transitional, and (c) turbulent flow.

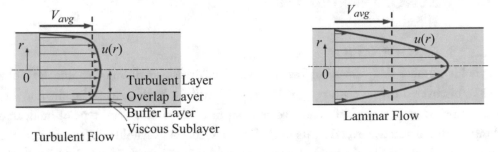

Figure 5.2: Development of velocity profile in turbulent and laminar flow.

Figure 5.3: Development of velocity profile in laminar flow.

5.1.2   OBJECTIVE

The purpose of this experiment is to determine the Reynolds number for laminar, transitional, and turbulent flow in a circular pipe. Furthermore, the aim of this lab is to visualize three flow types (Figure 5.1) and observe the development of velocity profile in a clear pipe (Figures 5.2 and 5.3).

5.1.3   PROCEDURE

1. Measure the tube diameter "D" to determine the cross-sectional area of the tube in the apparatus (Figure 5.4).

2. Obtain the temperature of water to determine density "ρ" and absolute viscosity "μ" from the textbook tables.

3. Use the to provide a supply to visualize the flow regime.

4. Adjust the control valve to provide laminar flow.

5. Allow the flow to stabilize and allow the velocity profile to develop a long the beginning of the tube. Take a picture of the velocity profile.

6. Obtain laminar flow and document a picture of the laminar flow.

7. Measure the volumetric flow rate using a graduated cylinder and stopwatch $Q = \frac{V}{t}$ (m³/sec).

8. Measure the flow rate three times to ensure the accuracy.

9. Repeat the experiment for transitional and turbulent flow.

10. Document observations of laminar flow by picture and support the observations for each flow type with computed Reynolds number in Tables 5.1, 5.2, and 5.3.

5.1.4   EQUIPMENT

1. Reynolds Number Apparatus

2. Blue Dye

3. Thermometer

4. Stopwatch

5. Graduated Cylinder

6. Camera

Figure 5.4: Reynolds number apparatus.

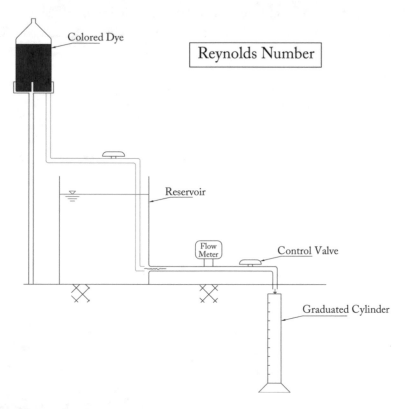

Figure 5.5: Reynolds number apparatus sketch.

5.1.5 EXPERIMENTAL DATA

Note: Use SI unites for this experiment.

Table 5.1: Experimental Data for Laminar Flow

Trial	Temp (°C)	Density (kg/m³)	Viscosity (m²/sec)	Volume of Water (mL)	Time (sec)	Flow Rate (m³/sec)	Velocity (m/sec)	R_e	Type of Flow
1									
2									
3									

Table 5.2: Experimental Data for Transitional Flow

Trial	Temp (°C)	Density (kg/m³)	Viscosity (m²/sec)	Volume of Water (mL)	Time (sec)	Flow Rate (m³/sec)	Velocity (m/sec)	R_e	Type of Flow
1									
2									
3									

Table 5.3: Experimental Data for Turbulent Flow

Trial	Temp (°C)	Density (kg/m³)	Viscosity (m²/sec)	Volume of Water (mL)	Time (sec)	Flow Rate (m³/sec)	Velocity (m/s)	R_e	Type of Flow
1									
2									
3									

5.1.6 LAB ASSIGNMENT

1. Complete three tables for the experimental data.

2. Calculate Reynolds number for the three flow types.

3. Provide supporting pictures for each flow regime.

5.2   EXPERIMENT: APPLICATION OF BERNOULLI'S EQUATION—MEASURE VELOCITY AND PRESSURE IN SIPHON

5.2.1   INTRODUCTION

A siphon is a u-shaped tube that is used to transfer liquid from a reservoir to a lower level due to atmospheric pressure in the beginning of the tube and gravity at the other end of the tube.

5.2.2   OBJECTIVE

The purpose of this lab is to determine the vacuum pressure at point B by applying Bernoulli's equation to the siphon.

5.2.3   SIPHON TOOLS

1. Two Buckets or Graduated Cylinders

2. Tube

3. Thermometer

4. Meter Stick

5.2.4   TEST SETUP

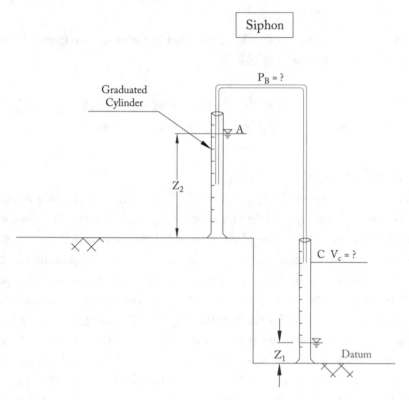

Figure 5.6: Siphon test set-up.

5.2.5   PROCEDURE

1. Set up a siphon at two heights and determine the difference between the two flow rates (Figure 5.6).

2. Measure the temperature with a thermometer to determine the density of water from the tables in the book.

3. Measure the height of point B from the assumed datum.

4. Use Bernoulli's Equation to calculate the required variables.

5.2.6   LAB ASSIGNMENT

1. State any assumptions for the calculations.

2. Calculate the vacuum pressure at B, velocity at point C, and the discharge in the tube for two different heights using Bernoulli's equation.

3. Do you think the head loss should be included in the energy equation? If yes, determine the rough estimate of head loss.

5.3   EXPERIMENT: MAJOR AND MINOR HEAD LOSS FOR TURBULENT FLOW IN A PIPE NETWORK

5.3.1   INTRODUCTION

In ideal flow, the total energy of the fluid within the control volume remains constant which is including fluid's pressure head, velocity head, and elevation head. This is called the conservation of energy principle. In reality, however, head loss is the major contributor in designing piping system due to friction between fluid and the surface, change in cross sectional area, fitting, bends, valves, etc. in which fluid travels upon. The head loss is mostly due to pressure drop through piping system. The total head loss can be calculated by simply dividing the pressure difference of a fluid between two points within a control volume by the specific weight of the fluid (Equation 5.2).

Darcy's Equation (Equation 5.4) states that the major head loss depends on the piping materials (friction factor) and diameter of the pipes, length in which fluid travels through, and velocity of the fluid. Minor head loss is also result of directly proportional to the velocity of the fluid. Friction factor "f" is directly proportional to two items: the ratio of absolute roughness for specific piping materials to the pipe's diameter and a fluid's Reynolds number" Re". Friction factor can be calculated Colebrook Equation (Equation 5.7) or Moody's graph. Sometimes the sum of minor losses is greater than major head loss. Thus, to design residential, municipal, or industrial piping systems, the impact of the total loss "h_L" (Equation 5.3) in the fluid pressure delivery is significant.

5.3.2   OBJECTIVES

The primary objective of this experiment is to determine the relationship between head loss of flow and the roughness of the material in the piping system in which flow was traveling through. In this experiment, the total head loss of flow between two points in a ductile iron, copper, and PVC pipe of equal length and diameter was determined by measuring the rate of flow and the difference in pressure of the fluid between two points in each pipe. Additionally, the average experimental roughness coefficient of each pipe material is calculated and compared to experimental values with theoretical values. The second objective of this experiment is to determine the relationship between total head loss of flow and the velocity at which a fluid was traveling through a PVC pipe.

Required Equations

$$h_L = \frac{\Delta P}{\gamma}.$$ (5.2)

where

h_L = Total Head Loss of Flow (ft),

ΔP = Pressure Differential (lb/ft^2),

γ = Specific Weight of Water (lb/ft^3),

$$h_L = f \frac{L}{D} \frac{V^2}{2g} + \sum K_L \frac{V^2}{2g}.$$ (5.3)

Major Head Loss—Darcy's Equation

$$h_{Major} = f \frac{L}{D} \frac{V^2}{2g}$$ (5.4)

Minor Head Loss

$$h_{Minor} = \sum K_L \frac{V^2}{2g},$$ (5.5)

where

h_L = Total Head Loss of Flow (ft)

L = Length of Pipe in which Head Loss is Measured (ft)

D = Diameter of Pipe (ft)

v = Velocity of Flow in Pipe (ft/sec)

g = Gravitational Acceleration Constant (ft/sec^2)

k_L = Minor Head Loss Coefficient (unitless)

$$Re = \frac{V_{ave} D}{v},$$ (5.6)

where

Re = Reynolds Number (unitless)

V_{avg} = Average Flow Velocity (ft/s)

D = Diameter of Pipe (ft)

v = Kinematic Viscosity of Water (ft^2/sec)

Colebrook Equation

$$\frac{1}{\sqrt{f}} = -2\log\left(\frac{\frac{\varepsilon}{D}}{3.7} + \frac{2.57}{Re\sqrt{f}}\right)$$ (5.7)

where

f = Experimental Friction Factor (unitless)

ε = Experimental Absolute Roughness Coefficient (ft),

D = Diameter of Pipe (ft)

Re = Reynolds Number (unitless)

5.3.3   EQUIPMENT

1. Piping System apparatus

2. Hydraulic Bench

3. Two Plastic Houses Connected to the Hydraulic Bench

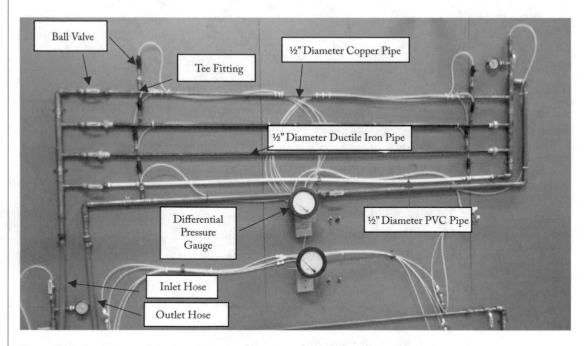

Figure 5.7: An image of the head loss apparatus used to perform experiments.

Figure 5.8: Flow is adjusted until a constant differential pressure reading is made by the gauge.

5.3.4   PROCEDURE

1. Connect the hydraulic bench to the inlet and outlet pipes of the apparatus by two plastic hoses.

2. Turn on the pump and the gate valve on the hydraulic bench to allow flow to enter the piping system until the flow reach to steady state.

3. Close all valves, but ductile iron pipe to allow water only flow through that pipe and to the differential pressure gauge.

4. Adjust flow using the gate valve until it has enough pressure such that a constant differential pressure reading can be made by the gauge.

5. Record the differential pressure gage.

6. The flow rate entering the hydraulic bench is determined by measuring the time it takes for flow to reach a certain volume in the flow piezometer.

7. Measure the water temperature by thermometer for flow properties such as specific weight in your calculation.

8. Close the ball valves for ductile iron pipe and open the ball valves to allow flow through copper pipe between the tee fittings. Use similar procedure to measure the flow rate and differential pressure reading.

9. Close the ball valves for copper pipe and open the ball valves to allow flow through PVC pipe between the tee fittings. Use similar procedure to measure the flow rate and differential pressure reading.

5.3.5   EXPERIMENTAL DATA

Note: Use BGS unites for this experiment.

Table5.4: Experimental Data

Pipe	Diameter (in)	Pipe Length (ft)	Roughness Coefficient (ft)	Flow Rate (L/sec)	Differential Pressure Gage (psi)
Ductile Iron	1/2	6	0.0027–0.005		
Copper	1/2	6	0.0000033–0.0000067		
PVC	1/2	6	0.0000233–0.0005		
Ductile Iron	3/4	6	0.0027–0.005		
Note: Record water temperature before starting the experiment.					

Table 5.5: Minor Head Loss Coefficients of Pipe Components

Pipe Component	Minor Head Loss Coefficient (KL)
Threaded Tee Fitting, Dividing Line Flow	0.9
Ball Valve, Fully Open	0.5

Table 5.6: Calculated Data for Ductile Iron Pipe in "Roughness" Experiment

	Flow Rate (L/sec)	Flow Rate (ft³/sec)	Velocity(ft/sec)	Total Head Loss (ft)	Major Head Loss (ft)	Experimental Friction Factor	Kinematic Viscosity of Water (ft²/sec)	Experimental Roughness Coefficient (ft)
Trial 1								
Trial 2								
Trial 3								
Trial 4								
Trial 5								

Table 5.7: Calculated Data for Copper Pipe in "Roughness" Experiment

	Flow Rate (L/sec)	Flow Rate (ft³/sec)	Velocity (ft/sec)	Total Head Loss (ft)	Major Head Loss (ft)	Experimental Friction Factor	Kinematic Viscosity of Water (ft²/sec)	Experimental Roughness Coefficient (ft)
Trial 1								
Trial 2								
Trial 3								
Trial 4								
Trial 5								

Table 5.8: Calculated Data for PVC Pipe in "Roughness" Experiment

	Flow Rate (L/sec)	Flow Rate (ft³/sec)	Velocity(ft/sec)	Total Head Loss (ft)	Major Head Loss (ft)	Experimental Friction Factor	Kinematic Viscosity of Water (ft²/sec)	Experimental Roughness Coefficient (ft)
Trial 1								
Trial 2								
Trial 3								
Trial 4								
Trial 5								

5.3.6   LAB ASSIGNMENT

1. Complete the tables and calculate experimental roughness.

2. Compare experimental roughness with theoretical value.

3. Calculate percentage error and source of errors.

4. Determine uncertainty.

5. Which pipe has the maximum head loss?

6. How do piping system's materials affect the minor and major head losses?

CHAPTER 6

External Flow

6.1 EXPERIMENT: MEASURE VELOCITY IN EXTERNAL FLOW USING PITOT TUBE

6.1.1 INTRODUCTION

Henri Pitot, a French hydraulic engineer, invented the Pitot tube to measure the flow velocity in the river (dynamic pressure) in 1732. He used a right-angled glass tube submerged in water with one end open to the atmosphere to determine fluid flow velocity.

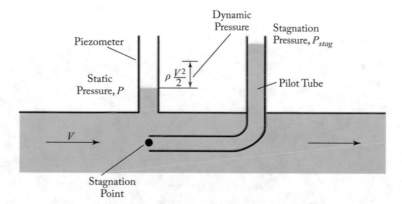

Figure 6.1: The static, dynamic, and stagnation pressure.

Required Equations

Pitot Tubes

$$V = \sqrt{\frac{2(P_0 - P)}{\rho}}$$ (6.1)

V = Velocity (m/s)

P_0 = Stagnation pressure

P = Static pressure

Bernoulli's Equation with Head Loss

$$\frac{P_a}{\gamma} + \frac{V_a^2}{2g} + Z_a = \frac{P_b}{\gamma} + \frac{V_b^2}{2g} + Z_b + h_L \tag{6.2}$$

6.1.2 OBJECTIVE

The purpose of this lab is to measure the energy balance by applying Bernoulli's equation to the upstream and downstream ends of a sluice gate in the flume.

6.1.3 APPARATUS

1. Pitot tube

2. Flume

3. Stick meter

4. Swoffer (velocity meter)

Figure 6.2: Open-channel flume.

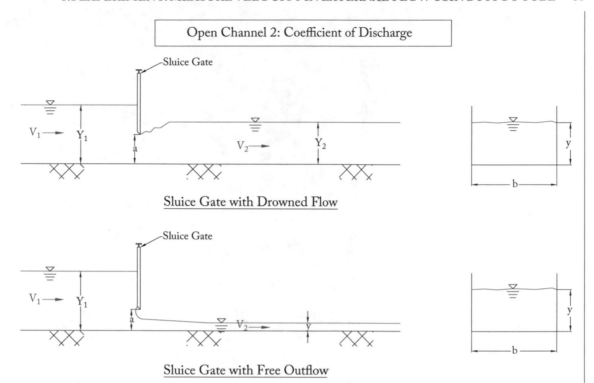

Figure 6.3: Sluice gate in the flume sketch.

6.1.4 PROCEDURE

1. Using a piezometer and Pitot tube, measure the pressure head and velocity head at the upstream and downstream ends.

2. Compare the experimental velocity with the Swoffer (Figure 6.4) and calculate the percent error.

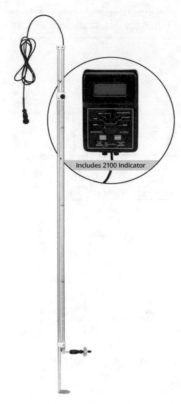

Figure 6.4: Swoffer Model 2100 Series and 3000 Series Open Stream Current Velocity Meters (https://www.forestry-suppliers.com/product_pages/products.php?mi=33981&itemnum=94160).

6.1.5   EXPERIMENTAL DATA

Table 6.1: Experimental Data for Upstream

Trial	Pressure Head P_1/γ (m)	Velocity Head $V_2^1/2\,g$ (m)	Elevation Head z_1 (m)
1			
2			
3			
Average			

Table 6.2: Experimental Data for Downstream

Trial	Pressure Head P_1/γ (m)	Velocity Head $V_2^1/2\,g$ (m)	Elevation Head z_1 (m)	Head Loss (m)
1				
2				
3				
Average				

Table 6.3: Actual Velocity Using Swoffer for Upstream and Downstream

Velocity Upstream Using Swoffer (m/sec)	Velocity Downstream Using Swoffer (m/s)
*Note: Use SI unites for this experiment.	

6.1.6   LAB ASSIGNMENT

1. Calculate the head loss between the upstream and downstream sides.

2. Compare the velocity head measured by the Swoffer and pitot tube.

3. Calculate the percent difference between these two values.

6.2   EXPERIMENT: COEFFICIENT OF DISCHARGE IN FLOW REGIME-WEIR

6.2.1   INTRODUCTION

Engineers use different shapes for the design of outlet structures to manipulate the discharge rate of a fluid. They often use circular shapes for an outfall; however, it is also common to see rectangular outfalls used for larger flows such as a dam spillway. These outlets impact the flow type resulting in either laminar, transitional, or turbulent flow.

Figure 6.5: Outfall shapes used in real world applications.

Required Equations

Rectangular Weir Flow

$$Q = C_d \frac{2}{3} B\sqrt{2g}H^{3/2} \tag{6.3}$$

C_d = Coefficient of discharge
B = Width of rectangular weir (m)
H = Height of water (head) above the weir crest (m)

Reynolds Number

$$Re = \frac{VR_h}{v} \tag{6.4}$$

V = Velocity $(\frac{m}{sec})$
R_h = Hydraulic radius (m)
v = Kinetic viscosity $(\frac{m^2}{sec})$

Hydraulic Radius

$$R_h = \frac{A_c}{P_w} \tag{6.5}$$

A_c = Cross-sectional area (m²)

P_w = Wetted Perimeter (m)

v = Kinetic viscosity $(\frac{m^2}{sec})$

6.2.2   OBJECTIVE

The purpose of this lab is to observe the effect of three weirs on the type of flow (laminar, transitional, or turbulent) produced by differently shaped outflow plates.

6.2.3   EQUIPMENT

1. Hydraulic Bench

2. Circular, 90° Wedge, and Square Plate

3. Hydraulic Bench

4. Piezometer

5. Meter Stick

6. Stopwatch

Figure 6.6: Hydraulic bench with three outlet plate apparati.

6.2.4   PROCEDURE

1. Measure the dimensions of the three plates.

2. Screw the rectangular plate to the weir carrier on the hydraulic bench.

3. Connect the pipe to the hydraulic bench nozzle as the water inlet.

4. Start the pump to provide steady, desirable flow by opening the adjustable valve in the hydraulic bench.

5. Measure the corresponding flow. Use a stopwatch and record the volume of water in the piezometer of the hydraulic bench in liters ($Q = \frac{V}{t}$, l/sec).

6. Record the wetted perimeter of the outlet plate.

7. Close the hydraulic bench valve and turn off the pump.

8. Repeat the procedure with the circular and triangular plates.

6.2.5   EXPERIMENTAL DATA

Table 6.4: Flow Results Obtained from the Experiment

Weir Shape	Cross Sectional Area (m²)	Wetted Perimeter (m)	Hydraulic Radius (m)	Density (m)	Velocity (m/sec)	Average Flow (m³/m)	Kinematic Viscosity (m/sec²)	Reynolds Number	Type of Flow
Circle									
Triangle									
Square									
*Note: Use SI unites for this experiment.									

Flow Classification in External Flow

Laminar flow: Re < 2300

Transitional flow: 2300 < Re < 4000

Turbulent flow: Re > 4000

6.2.6   LAB ASSIGNMENT

1. Complete Table 6.4.

2. Calculate the coefficient of discharge for the rectangular weir.

3. Discuss what you observed and learned from this experiment as a group and write a paragraph of your discussion.

6.3   EXPERIMENT: OPEN-CHANNEL FLOW-SPECIFIC ENERGY-SUBCRITICAL, CRITICAL AND SUPERCRITICAL FLOW

6.3.1   INTRODUCTION

Open-channel flow refers to the flow of liquids in channels open to the atmosphere or in partially filled conduits and is characterized by the presence of a liquid–gas interface called the free surface, examplesvof which include streams, rivers, and culverts not flowing full.

Required Equations

Reynold's Number for Laminar and Turbulent Flow in Channels

$$R_e = \frac{VR_h}{v} = \frac{\rho VR_h}{\mu} \tag{6.6}$$

R_h = Hydraulic radius (m)
V = Velocity ($\frac{m}{sec}$)
$v = \mu/\rho$ = Kinematic viscosity ($\frac{m^2}{sec}$)
ρ = Density ($\frac{kg}{m^3}$)
μ = Absolute viscosity ($\frac{m^2}{sec}$)

Note: For an open channel, the flow is determined as follows:

Laminar flow: Re ≤500

Transitional flow: 500 < Re < 2500

Turbulent flow: Re > 2500

Hydraulic Radius

$$R_h = \frac{A_c}{P} \tag{6.7}$$

A_c = Cross-sectional area of flow (m^2)
P = Wetted perimeter (m)

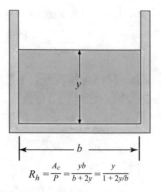

$$R_h = \frac{A_c}{P} = \frac{yb}{b+2y} = \frac{y}{1+2y/b}$$

Figure 6.7: Hydraulic radius of a rectangle channel.

Hydraulic Diameter

$$D_h = 4R_h \tag{6.8}$$

Froude Number

$$F_r = \frac{V}{\sqrt{gy}} \tag{6.9}$$

V = Velocity ($\frac{m}{sec}$)

g = Gravitational Acceleration Constant ($\frac{m}{sec^2}$)

y = Channel depth (m)

Note: The Froude number is an important parameter that governs the character of flow in open channels. The flow is classified as:

Fr<1, Subcritical or tranquil flow,

Fr=1, Critical flow,

Fr>1, Supercritical or rapid flow.

Critical Depth

$$y_c = \frac{Q^2}{gA^2} \tag{6.10}$$

Q = Flow rate ($\frac{m^3}{sec}$)

g = Gravitational Acceleration Constant ($\frac{m}{sec^2}$)

Ac = Cross-sectional area of flow (m²)

Note: Another parameter to classify the type of flow in open channel is the critical depth. The flow is classified as follows:

$y > y_c$, Subcritical flow,

$y < y_c$, Supercritical flow.

Specific Energy

$$E_s = y + \frac{V^2}{2g} = y + \frac{Q^2}{2gb^2y^2}$$
(6.11)

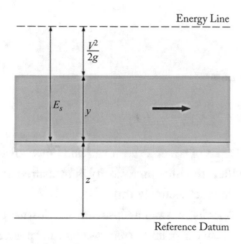

Figure 6.8: Specific energy variables.

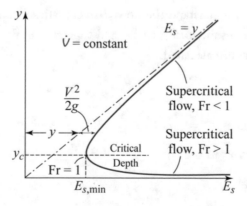

Figure 6.9: Variation of specific energy with depth for a specified flow rate.

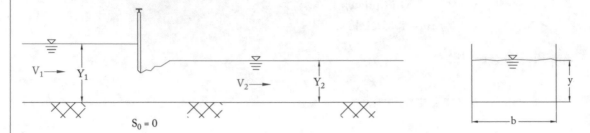

Figure 6.10: Flume apparatus.

Energy Equation and Head Loss

$$Z_1 + y_1 + \frac{V_1^2}{2g} = Z_2 + y_2 + \frac{V_2^2}{2g} + h_L \tag{6.12}$$

y_1 and y_2 = Depth of water at upstream and downstream flow, respectively (m)

Z_1 and Z_2 = Elevation that the velocity is measured at upstream and downstream flow, respectively (m)

V_1 and V_2 = Velocity of water at upstream and downstream flow, respectively (m/s)

h_L = Head loss between point 1 at the upstream and point 2 at the downstream end (m)

6.3.2   OBJECTIVE

The purpose of this lab is to understand the flow characteristics in open channels. To determine fluid's classification, including subcritical, critical, and supercritical flow, the head loss and critical depth in open-channel must be calculated.

6.3.3   EQUIPMENT

1. Flow channel

2. Pump to supply water to the flume

3. Swoffer or pitot tube

4. Meter stick

6.3.4   PROCEDURE

1. Record the geometry of the rectangular open channel.

2. Connect the delivery hose from the valve to the suction part of the pump.

3. Open the valve to prime the pump.

4. Adjust the sluice gate for proper flow upstream and downstream.

5. Start the pump to allow for water to enter the flume. Begin recording data once the flow rate remains constant.

6. Record the depth of water and velocity at the upstream and downstream ends.

7. Repeat these measurements five times to ensure accuracy of the data.

6.3.5 EXPERIMENTAL DATA

*Note: Use SI unites for this experiment.

Table 6.5: Experimental Data for Reynolds Number in Open-Channel Flow-Upstream

Trial	$V(m/s)$	R_h	D_h	$T(°C)$	$v(m^2/sec)$	R_e	Laminar	Transitional	Turbulent
1									
2									
3									
4									
5									
Average									

Table 6.6: Experimental Data for Reynolds Number in Open-Channel Flow-Downstream

Trial	$V(m/s)$	R_h	D_h	$T(°C)$	$v(m^2/sec)$	R_e	Laminar	Transitional	Turbulent
1									
2									
3									
4									
5									
Average									

Table 6.7: Experimental Data for Froude Number in Open-Channel Flow-Upstream

Trial	V(m/sec)	g(m/sec^2)	y(m)	Q(m^3/sec)	A(m^2)	v(m^2/sec)	Fr	y_c(m)	Subcritical	Critical	Supercritical
1											
2											
3											
4											
5											
Average											

Table 6.8: Experimental Data for Froude Number in Open-Channel Flow-Downstream

Trial	V(m/sec)	g(m/sec^2)	y(m)	Q(m^3/sec)	A(m^2)	v(m^2/sec)	Fr	y_c(m)	Subcritical	Critical	Supercritical
1											
2											
3											
4											
5											
Average											

Table 6.9: Experimental Data for Head Loss in Open-Channel Flow

Trial	Z_1(m)	y_1(m)	$V_1^2/2g$(m)	Z_2(m)	y_2(m)	$V_2^2/2g$(m)	h_L(m)
1							
2							
3							
4							
5							
Average							

Table 6.10: Experimental Data for Specific Energy and Froude Number in Open-Channel Flow

Trial	v(m²/sec)	Fr	y_c(m)	Subcritical	Critical	Supercritical
1						
2						
3						
4						
5						
Average						

6.3.6   LAB ASSIGNMENT

1. Complete the tables above and make a sketch of the sluice gate, EGL, HGL, and calculate yc.

2. Discuss what you observed and learned from this experiment as a group and write a paragraph of your discussion.

6.4   EXPERIMENT: HYDRAULIC JUMP AND HEAD LOSS IN OPEN-CHANNEL FLOW

6.4.1   INTRODUCTION

Rapidly varied flow occurs when there is a sudden change in flow, such as an abrupt change in the cross-section of a channel. Rapidly varied flows are typically complicated because they involve significant multidimensional and transient effects, backflows, and flow separation. Therefore, rapidly varied flows are usually studied experimentally or numerically. It is still possible to analyze some rapidly varied flows using the one-dimensional flow approximation with reasonable accuracy. The fraction of energy dissipation ranges from just a few percent for weak hydraulic jumps ($Fr_1 < 2$) to 85° for strong jumps ($Fr_1 > 9$).

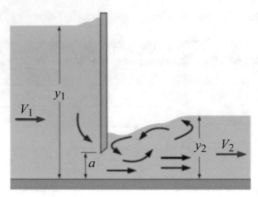

Figure 6.11: Example of a hydraulic jump.

Required Equations

Energy Dissipation Ratio

$$\text{Energy dissipation ratio} = \frac{h_L}{E_{s1}} = \frac{h_L}{y_1 + \frac{v_1^2}{2g}} = \frac{h_L}{y_1 \left(1 + \frac{Fr_1^2}{2}\right)} \tag{6.13}$$

h_L = Head Loss (m)
E_{s1} = Energy Before the Hydraulic Jump (m)
y_1 = Depth of water at upstream flow (m)
V_1 = Velocity of water at upstream flow ($\frac{m}{sec}$)

Head Loss in a Hydraulic Jump

$$h_L = y_1 - y_2 + \frac{y_1 Fr_1^2}{2}\left(1 - \frac{y_1^2}{y_2^2}\right) \tag{6.14}$$

y_1 = Upstream Height (m)
y_2 = Downstream flow Height (m)
F_{r1} = Freud's Number Before the Hydraulic Jump

Depth Ratio for a Hydraulic Jump

$$\frac{y_2}{y_1} = 0.5\left(-1 + \sqrt{1 + 8F_{r1}^2}\right) \tag{6.15}$$

y_1 = Upstream Height (m)
y_2 = Downstream Height (m)

F_{r1} = Freud's Number Before the Hydraulic Jump

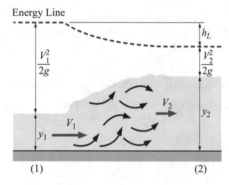

Figure 6.12: Hydraulic jump variables.

6.4.2   OBJECTIVE

The purpose of this lab is to determine the head loss in a hydraulic jump and the energy dissipation ratio associated with this change.

6.4.3   EQUIPMENT

1. Flow channel

2. Pump to supply water to the flume

3. Swoffer or pitot tube

4. Meter stick

6.4.4   PROCEDURE

1. Record the geometry of the rectangular open channel.

2. Record the height of the sluice gate opening.

3. Connect the delivery hose from valve to the suction part of the pump.

4. Open the valve to prime the pump.

5. Adjust the sluice gate for proper flow upstream and downstream.

6. Start the pump to allow water to enter the flume. Begin recording data once the flow rate remains constant.

7. Record the depth of water upstream and downstream. Determine the velocity using the Swoffer.

8. Determine the theoretical coefficient of discharge "C_d" value from the equation below and compare this with the experimental value of coefficient of discharge (Figure 6.13).

$$Q = C_d \, ba\sqrt{2gy_1} \qquad\qquad (6.16)$$

Q = Flow rate (m³/sec)
C_d = Coefficient of discharge (unitless)
y_1 = Depth of flow at upstream (m)
a = Sluice gate opening height (m)
b = Width of open channel (m)

9. Repeat the measurement five times to ensure the accuracy of the data.

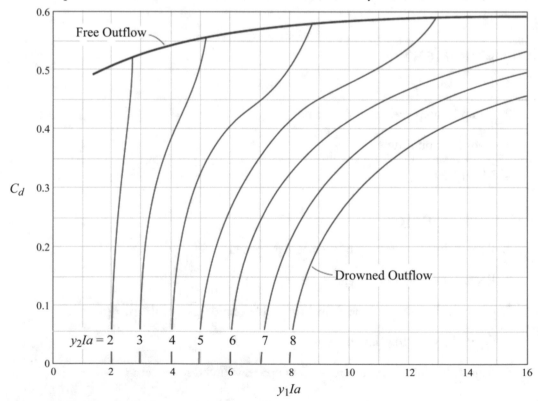

Figure 6.13: Discharge coefficients for drowned and free discharge from underflow gates.

6.4.5   EXPERIMENTAL DATA

Table 6.11: Experimental Data and Coefficient of Discharge

V_1(m/s)	y_1(m)	y_2(m)	Fr Number	h_L(m)	Dissipation Ratio
*Note: Use SI unites for this experiment.					

6.4.6   LAB ASSIGNMENT

1. Determine Froude number and flow classification at upstream and downstream of sluice gate.

2. Calculate the energy dissipation ration and the percentage of energy dissipation.

3. Determine the head loss and impact of sluice gate height on the energy dissipation.

4. Discuss what you observed and learned from this experiment as a group and write a paragraph of your discussion.

References

Ahmari, H. and Kabir, S. M. I. (2019). *Applied Fluid Mechanics Lab Manual*. Mavs Open Press. 36

Alastal, K. M. and Mousa, M. Y. (2015). Fluid Mechanics and Hydraulics Lab Manual: http://site.iugaza.edu.ps/mymousa/files/Fluid-Mechanics-and-Hydraulics-Lab-Manual-2015-.pdf.

Cimbala, J. and Cengel, Y. A. (2018). *Fluid Mechanics: Fundamentals and Applications*, 4th edition. McGraw-Hill Higher Education. 16, 23, 65

Lab report format, Department of Mechanical Engineering, School of Engineering, Boston University website: http://www.bu.edu/eng/departments/me/general-resources-students/current-undergraduate-students/lab-report-format/.

State Institute of Technical Teachers Training and Research, Kalamassery, Lab Manual for Fluid Mechanics Lab: http://www.sitttrkerala.ac.in/misc/LabManual/FML.pdf.

Author Biography

Robabeh Jazaei is an Assistant Professor of Civil Engineering in the Department of Physics and Engineering at Slippery Rock University of Pennsylvania. She received her Ph.D. in Civil and Environmental Engineering from the University of Nevada, Las Vegas.

Dr. Jazaei has years of professional and academic experience. She taught a broad range of courses and laboratories with the focus of solid and fluid mechanics, as well as structural design in several universities. She is well known for practicing pedagogy with illustrative approach to simulate the interest in engineering courses and facilitate students' learning. Dr. Jazaei taught fluid mechanics in 2002 for the first time and since then she has been inspired to provide graphical guidance to help students to enjoy the fluid mechanics' laboratory. From 2018–2020, she has not only lectured fluid mechanics continuously but also worked closely with civil engineering students in 14 laboratories at the University of Wisconsin-Platteville. During these years, Dr. Jazaei realized the need to help students perform fluid mechanics experiments and write an effective technical report.

Dr. Jazaei also served as an adviser for many senior design and research teams. Her research interest has focused on a broad spectrum of the numerical and experimental analysis of structures, concrete failure mechanisms, and nanomaterials characterization in cementitious composites. She received several academic and professional certifications for design, construction management, and professional development.

Dr. Jazaei is a voting member of American Concrete Institute (ACI) Committee 241-A01 (The Application and Implementation of Nano-Engineered Concrete), member of American Society of Civil Engineers (ASCE), American Society of Mechanical Engineering (ASME), and American Society of Heating, Refrigerating and Air-Conditioning Engineers (ASHRAE).

Printed in the United States
by Baker & Taylor Publisher Services